PUBLIC PERCEPTION OF BIOSOLIDS RECYCLING: DEVELOPING PUBLIC PARTICIPATION AND EARNING TRUST

by:

Ned Beecher
New England Biosolids and Residuals Association

Barry Connell
Center for Environmental Communications

Eliot Epstein
Jennifer Filtz
E & A Environmental Consultants

Nora Goldstein
BioCycle

Maile Lono
Northwest Biosolids Management Association

2004

WERF

The Water Environment Research Foundation, a not-for-profit organization, funds and manages water quality research for its subscribers through a diverse public-private partnership between municipal utilities, corporations, academia, industry, and the federal government. WERF subscribers include municipal and regional water and wastewater utilities, industrial corporations, environmental engineering firms, and others that share a commitment to cost-effective water quality solutions. WERF is dedicated to advancing science and technology addressing water quality issues as they impact water resources, the atmosphere, the lands, and quality of life.

For more information, contact:
Water Environment Research Foundation
635 Slaters Lane, Suite 300
Alexandria, VA 22314-1177
Tel: (703) 684-2470
Fax: (703) 299-0742
www.werf.org
werf@werf.org

This report was co-published by the following organizations. For non-subscriber sales information, contact:

Water Environment Federation
601 Wythe Street
Alexandria, VA 22314-1994
Tel: (800) 666-0206
Tel: (703) 684-2452
Fax: (703) 684-2492
www.wef.org
pubs@wef.org

IWA Publishing
Alliance House, 12 Caxton Street
London SW1H 0QS, United Kingdom
Tel: +44 (0) 20 7654 5500
Fax: +44 (0) 20 7654 5555
www.iwapublishing.com
publications@iwap.co.uk

Library of Congress Catalog Card Number: 2004115485

WEF ISBN:1-57278-228-5 IWAP ISBN: 1-84339-710-2

This report was prepared by the organization(s) named below as an account of work sponsored by the Water Environment Research Foundation (WERF). Neither WERF, members of WERF, the organization(s) named below, nor any person acting on their behalf: (a) makes any warranty, express or implied, with respect to the use of any information, apparatus, method, or process disclosed in this report or that such use may not infringe on privately owned rights; or (b) assumes any liabilities with respect to the use of, or for damages resulting from the use of, any information, apparatus, method, or process disclosed in this report.

New England Biosolids and Residuals Association

This document was reviewed by a panel of independent experts selected by WERF. Mention of trade names or commercial products does not constitute WERF endorsement or recommendations for use. Similarly, omission of products or trade names indicates nothing concerning WERF's positions regarding product effectiveness or applicability.

ACKNOWLEDGMENTS

Report Preparation

Principal Investigator:
Ned Beecher, MS, Executive Director
New England Biosolids and Residuals Association (NEBRA)

Project Team:
Barry Connell, Executive Director
Center for Environmental Communications

Eliot Epstein, P.E., Ph.D.
Jennifer Filtz
E & A Environmental Consultants

Nora Goldstein, Executive Editor
BioCycle

Maile Lono, Manager
Northwest Biosolids Management Association (NBMA)

Additional Case Study Authors:
Joel Alpert, Ph.D.
Compost and Technology Solutions, Inc.

Ken Lee, P.E.
Greater Vancouver Regional District

Peggy Leonard and Roberta King
King County Department of Natural Resources

Dan Thompson
City of Everett, WA

Project Subcommittee

Stephen W.C. Constable, P.E., DEE
DuPont Engineering

Stephen Frank, APR
Metro Wastewater Reclamation District, Denver, CO

Peter Nowak, Ph.D.
University of Wisconsin, Madison

Maureen McReynolds, Ph.D.
Water & Wastewater Utility, Austin, TX

Biosolids Stakeholder Review Panel

Note: Reviewers listed below do not necessarily agree with all parts or findings of this report. All interpretations, findings, and recommendations are the sole responsibility of the authors.

Robert E. Adamski, P.E., *Gannett Fleming*
Robert Bastian, *U.S. Environmental Protection Agency*
Robert Brobst, *U.S. Environmental Protection Agency, Region 8*
John F. Donovan, P.E., *CDM*
Jane Forste, *Jane Forste Associates*
Rob Greenwood, *Ross & Associates Environmental Consulting Ltd.*
Sam Hadeed, *National Biosolids Partnership*
Ellen Harrison, *Cornell Waste Management Institute*
Troy Hartley, *Northeast Consortium, University of New Hampshire*
Anthony Ho, *Ontario Ministry of the Environment*
Uta Krogmann, Ph.D., *Rutgers University*
Peggy Leonard, *King County Biosolids Program*
Trille Mendenhall, *Charlotte-Mecklenburg Utilities, Charlotte, NC*
Michael Moore, *Orange County Sanitation District*
Bob O'Dette, *Synagro*
David Oerke, P.E., *RTW Engineers*
Janice Patterson, *Ontario Ministry of Agriculture and Food & CH2M Hill*
Ronald Sieger, P.E., DEE, *CH2M Hill*
Jody Slagle, *Austin Water/Wastewater Department*
Dan Thompson, *NBMA, City of Everett and City of Tacoma, WA*
William Toffey, *Philadelphia Water Department*
Denise Uzupis, *Pennsylvania Department of Environmental Protection*

Social Science Consultants

Patrick Field, MS, *Vice-President, Consensus Building Institute and Associate Director, MIT-Harvard Public Disputes Program*
Merrick Hoben, *The Consensus Building Institute*
Douglas McKenzie-Mohr, Ph.D., *Professor of Psychology, St. Thomas University, New Brunswick*
Elaine Vaughan, Ph.D., *Professor of Social Ecology, University of California at Irvine*
Lawrence Susskind, Ph.D., *Ford Professor of Urban and Environmental Planning, Massachusetts Institute of Technology and President, The Consensus Building Institute*

Water Environment Research Foundation Staff

Director of Research: Daniel M. Woltering, Ph.D.
Project Manager: Bonie Bailey

AUTHOR INFORMATION

Ned Beecher holds a B.A. in Geology from Amherst College and an M.S. in Resource Management from Antioch University. After working in environmental education and residuals management, he became the first director of the New England Biosolids and Residuals Association (NEBRA) in 1998, where he administers membership services, provides an information network hub, and assists with public outreach and participation efforts regarding the recycling of biosolids and other organic residuals. He is the lead author of the 2001 NEBRA report *Saving Soil: Biosolids Recycling in New England.* His work on public perceptions and participation has led to continued consulting with WERF in developing WERF's "Public Partnering" initiative. Through NEBRA, and in conjunction with project associates, he offers workshops and consulting on public participation in biosolids management programs. He lives in rural New Hampshire with wife, Christine Clyne, and two children.

Barry Connell is Executive Director of the non-profit Center for Environmental Communications. The Center conducts research and educational projects in waste management; it works to limit exposure to products or wastes that may degrade human health or environmental quality; and promotes public policies and private investment in services or technologies that maximize resource conservation and pollution prevention. Mr. Connell is also currently a middle school teacher near his home in Newburyport, MA.

Eliot Epstein, Ph.D., was a principal in the Massachusetts firm E & A Environmental Consultants . He is currently with Tetratech, an adjunct professor at the Boston University School of Public Health, and co-editor of a new journal of biosolids management science. He resides southeast of Boston.

Jennifer Filtz, B.S., was with E & A Environmental Consultants during her work on this project.

Nora Goldstein is Executive Editor of *BioCycle, Journal of Composting and Organics Recycling*, published by The JG Press, Inc. in Emmaus, Pennsylvania. In addition to having written over a 100 articles related to biosolids recycling, she has conducted a number of surveys on biosolids management practices nationwide. Goldstein has participated on several advisory committees related to development of the Environmental Management System program administered by the National Biosolids Partnership, and was a co-author/co-editor of the *National Manual of Good Practice for Biosolids,* prepared under contract with the Water Environment Research Foundation. In 2003-2004, she is on the project team assisting WERF in its new "Public Partnering" initiative.

Maile Lono holds a B.A. in Community & Environmental Planning from the University of Washington and is pursuing a Masters in Not-for-Profit Leadership at Seattle University. Maile has been the Manager at Northwest Biosolids Management Association (NBMA) since 1999. The NBMA began activities in 1988 and was incorporated as a non-profit association in 1994 with the mission to advance environmentally sound biosolids management. The NBMA focuses it's goals and activities around providing a regional voice in the Northwest for biosolids management by offering networking and educational opportunities for members, regulators, private companies and the public; participating in the development of local, state, and federal biosolids regulations; and dedicating funding towards ongoing biosolids research. The NBMA has won awards from the U.S. EPA and Washington Department of Ecology for their leadership in biosolids management.

ABSTRACT AND BENEFITS

Abstract:

This report provides an overview of the current public perception and public acceptance of biosolids recycling in North America. It provides conceptual models for understanding the development of current public perceptions, including the influence of many factors--technical issues, typical human responses, communications, information, politics, and social context.

The report includes an annotated bibliography and narrative review of recent literature regarding public perceptions and public acceptance of biosolids recycling. Fourteen succinct case studies provide more detailed information on representative biosolids recycling program experiences. There are examples of biosolids management programs that have failed because of public controversy, as well as programs that have worked well, with community support. The latter programs have several things in common: they are well-run operations following best management practices; their benefits to communities and the environment are obvious; they communicate well with—and involve—stakeholders; most were introduced to communities through a respectful, mutually-beneficial communication process; and they had strong organizational commitment behind these public outreach efforts.

In order to assess current public knowledge and perception regarding biosolids recycling, a survey was conducted for this project by the University of New Hampshire Survey Center. The survey results include:

- Knowledge of "biosolids" is limited, but growing (14% can define it fairly accurately)
- Support for wastewater treatment is very high (93%).
- People are initially uneasy about biosolids recycling and cannot accurately assess its risks.
- The choice of words makes a difference—using "sewage sludge" instead of "biosolids" creates more negative response.
- Most people think that biosolids recycling is a relatively small risk to them.
- Although they clearly expressed support of the concept of recycling biosolids, respondents indicated that the strongest argument against biosolids recycling is "not enough is known," indicating that they are uncomfortable with their own personal lack of information and/or what they perceive to be an overall lack of information on this topic.

Because there is growing entrenched conflict regarding the benefits and risks of biosolids recycling, one part of this project has been to promote greater public participation, including alternative dispute resolution and consensus building, within the field of biosolids recycling. On April 11, 2002, the project held a day-long workshop on consensus building, led by Dr. Lawrence Susskind of the Harvard/MIT dispute resolution program. Biosolids management leaders from across the continent attended. A summary of the workshop is included in the report.

Through increased public participation, biosolids managers will increase the likelihood of reducing the risk of sudden costly setbacks. The report provides some strategies for evaluating public perception of a biosolids program and for selecting appropriate tools for public outreach and participation.

Benefits:

- Informs biosolids managers and other stakeholders about the current range and tenor of public perceptions of biosolids recycling in North America—including information from the largest nationwide survey on biosolids public perception.
- Provides understanding regarding how the public's perceptions of biosolids recycling are formed and what influences those perceptions.
- Provides biosolids managers and other stakeholders with information about the impacts on public perception and participation of various biosolids management program communications, actions, and other factors.
- Provides biosolids managers with background information and understanding about social science research regarding similar environmental and public policy issues, so that biosolids managers may benefit from the findings and recommendations of that body of research.
- Provides recommendations as to how biosolids management programs can more effectively communicate with and involve the public and key stakeholders such as site and facility neighbors, local communities, public interest groups, environmental groups, etc.
- Provides biosolids management programs with recommended resources, strategies, and tools for developing enhanced two-way communications, greater public participation, and the trust of communities in which they operate.

Keywords: biosolids management, sewage sludge, sludge, public acceptance, public education, public opinion survey, consensus building, environmental education, stakeholder

TABLE OF CONTENTS

The following sections are available on the accompanying CD-ROM:

LIST OF TABLES

LIST OF FIGURES

EXECUTIVE SUMMARY

The ultimate goal of a biosolids recycling program's outreach, education, public relations, and public participation efforts should be to develop "public relationships" with the variety of stakeholders who may be involved with and/or potentially impacted by a biosolids recycling program. Developing relationships through listening to, and having dialogue with, all stakeholders leads to increased trust and respect, as well as increased public understanding of biosolids recycling. In most instances, this relationship-building approach also will build better biosolids management programs that are supported widely by a diverse public. This report provides recommendations for how biosolids managers can increase constructive relationships with communities, neighbors, and other public stakeholders.

To develop their recommendations, investigators collected about 1,000 references from peer-reviewed, published and gray literature regarding public perceptions and interactions with biosolids recycling programs and social science literature on communications and public perceptions and participation. A searchable electronic, annotated bibliography is **included on the enclosed CD-ROM.**

The team also developed 14 case studies that document and analyze a variety of experiences with public perceptions and public involvement in biosolids recycling programs. The details of the case studies are **included on the enclosed CD-ROM.**

Based on the literature review and case studies, the investigators developed the first-ever, large-scale national survey regarding public perceptions of biosolids recycling. The University of New Hampshire Survey Center conducted the survey by telephone in early 2002, collecting opinions from more than a thousand U.S. residents. The details of this survey are **included on the enclosed CD-ROM.**

Finally, investigators held a one-day workshop in the Boston area in April 2002 in order to bring additional expertise from the social sciences to the project. At this workshop, the project team and advisors from around the continent learned more about how public perception develops, what affects the public's understanding of and response to a proposed project, and how biosolids managers might better interact with the public to gain greater trust and improve their programs. Details from this workshop are **included on the enclosed CD-ROM.**

The results of these four steps were brought together into this document, Developing Public Participation and Earning Trust.

Literature Review

The literature review was conducted to learn what has been written about biosolids recycling by those in the biosolids management field as well as interested persons outside the field. The included literature pertained to 1) public knowledge, perceptions, opinions, and behavior about biosolids recycling, and 2) the most significant and summary literature regarding public knowledge, perceptions, opinions, and behavior around related environmental issues (e.g., siting of facilities, solid waste management, energy, and new technologies). A wide variety of perspectives on environmental issues and biosolids recycling were included.

Case Studies

In consultation with a group of biosolids recycling stakeholders convened for this project, the investigating team chose to develop case studies about public acceptance issues and efforts at fourteen different biosolids recycling programs around North America. The experiences described in these case studies range from programs that have developed considerable public support by developing effective communications, trust, and long-term relationships with key "gatekeepers" (e.g., King County, WA) to programs that were stopped by public outrage (e.g., Kern County, CA). The case studies provided additional insights into how public acceptance issues noted in the literature review play out as biosolids recycling programs interact with the public.

The 2002 Biosolids Knowledge and Perception Survey

Based on the understanding developed from studying the literature and case studies, the investigating team developed a nationwide survey questionnaire, with assistance from the University of New Hampshire Survey Center. The Survey Center conducted the formal survey by telephone utilizing a nationwide sample that, in the end, totaled 1,069 individuals. Those surveyed were selected at random from the nationwide pool of home owners and house renters—people thought more likely to know something about sewage treatment and agriculture/gardening. Significant findings include:

- Knowledge of the term "biosolids" is limited (only 3% can accurately define "biosolids" and another 11% had a fairly good idea of what "biosolids" are).
- Support for wastewater treatment is very high (93%).
- Regional differences are minimal regarding the level of public knowledge and perceptions of biosolids recycling.
- People are uneasy and have questions about biosolids recycling.
- The choice of words makes a significant difference: using "sewage sludge" instead of "biosolids" creates a marked drop in positive response to a neighbor's hypothetical use of the material.
- If they need to learn more about biosolids, people want more information about many different aspects of biosolids recycling.
- Faced with a hypothetical situation in which biosolids are used by a neighbor, people say they would turn to and trust friends and neighbors, government agencies, and academic researchers for initial information.
- Faced with a hypothetical situation in which biosolids are used by a neighbor, people say they would trust government and others who appear knowledgeable and objective; they strongly distrust those with a profit motive.
- People favor constructive uses of biosolids (creating energy, recycling nutrients).
- Some factors increase people's concern about biosolids (if it includes an industrial waste source or is from a large city), others decrease their concern (if it is certified annually, if they are contacted prior to its use, if it is supervised locally).
- Strong arguments in favor of biosolids recycling are: Biosolids recycling returns nutrients to the soil, and recycling biosolids disposes of a necessary waste.

- Most people think that biosolids recycling is a relatively small risk to them.
- Although they clearly expressed support for the concept of recycling biosolids, respondents indicated that the strongest argument against biosolids recycling is "not enough is known," indicating that they are uncomfortable with their own lack of information and/or what they perceive to be an overall lack of information on this topic.

The 2002 Biosolids Public Knowledge and Perception Survey provides biosolids managers and other stakeholders with information about many factors and actions that will impact their chances of building trust, improving communications, and building public support for biosolids recycling programs.

Workshops on Public Participation

One important aspect of developing stronger public input and participation in biosolids management is training. How to interact with the public is not a common training topic in the wastewater management and engineering fields. Therefore, the project sponsored a one-day workshop to provide initial training on conflict resolution and the "mutual gains approach." In attendance were project advisors, biosolids managers, and other stakeholders from around North America. The summary from that workshop appears in Appendix D. Lessons learned from this and other workshops conducted subsequently at Water Environment Federation conferences have been incorporated into the concepts and recommendations included in this report.

Recommendations for Working with the Public

Managing sewage sludge is an important public function in modern society. Sewage sludge and biosolids managers should recognize that:

- They do not have to decide all by themselves how to manage sewage sludge and/or biosolids;
- If they want to ensure that the decisions being made are the best possible—and the most publicly supportable—they should get input from a variety of stakeholders (i.e., they should strive for and insist on public participation);
- If they are going to involve the public in meaningful, constructive ways, they must build "public relationships" with individual stakeholders; and
- Building public relationships requires attention to the relationship-building factors of trust, fairness, honest and useful information, two-way communications, and organizational commitment and motivation to do things right for most stakeholders.

In short, all public outreach, communications, and participation efforts and actions, as well as many technical decisions regarding biosolids management, should be developed and implemented with this question in mind:

"What will be the impact of my next action on our program's relationship(s) with key stakeholders?"

The report discusses twelve steps for developing and implementing a public outreach and participation program (see Table I-1). Many of the recommendations are not new, but their consistent implementation by biosolids managers has not yet been realized.

Biosolids managers are encouraged to pay attention to social aspects of their programs. An overarching guiding principle is to say things and take actions that build constructive long-term public relationships. Building positive public relationships hinges on the following principles, which have become central WERF themes identified from social science literature and experience through this project and a related one. Think of public relationships as an umbrella held up by:

- trust;
- fairness;
- quality of information;
- excellent communications; and
- your organizational commitment to public outreach, dialogue, and participation.

In addition, the following factors are important for biosolids managers to keep in mind:

- Listen to stakeholders; encourage their input; never discount or scoff at questions or concerns—most questions are reasonable to those raising them.
- Demonstrate that you care about protecting the environment. Ensure continual improvement in practices, and consider how you can remove factors that contribute to nuisances or public outrage (e.g., size and site projects appropriately). An Environmental Management System can help institutionalize continual improvement.
- Ensure independent oversight, either by a state or county agency or by creating independent oversight at the local level: it builds public trust, credibility, and confidence in your program.
- Give the public time to learn; big trouble comes when people feel pressured by looming deadlines. This may mean communicating extensively and getting public input long before any required public hearing or application deadlines, so that people are already well-informed when the deadlines arrive.
- Engage stakeholders as early in the process as possible: since biosolids recycling is a public function, interested citizens have a right and a responsibility to constructively assist the biosolids manager. Engaging constructive stakeholders may be one of your biggest challenges—so start working on it as soon as possible.
- Focus on building long-term relationships: Remember to ask yourself "What will be the impact of my next action on our program's relationship(s) with key stakeholders?"

Remember, you are not giving up responsibility or control by reaching out and involving stakeholders. You are still ultimately responsible, but you will have greater confidence (and your program will be less vulnerable) knowing that different stakeholders know about, have been involved in, and accept what you are doing. The information contained in this report will help you develop constructive relationships with communities, neighbors, and other public stakeholders.

INTRODUCTION

Project Overview

The nation's dependence on effective wastewater treatment requires an increasing capacity for the management of sewage sludge and biosolids. The amount of municipal sewage sludge generated annually continues to rise. The U.S. Environmental Protection Agency (U.S. EPA) estimates that more than seven (7) million dry tons of sewage sludge are generated each year, compared with four and one half (4.5) million dry tons in 1972, and that U.S. municipalities will produce more than eight (8) million dry tons annually by the year 2010. Approximately 60% of the sewage sludge produced today is applied to land as biosolids fertilizers or soil amendments, an increase from 33% in 1988. This trend, in conjunction with the increasing amount of biosolids produced, means that more land application sites are being proposed to meet a growing demand.

From September of 2000 through 2003, investigators under contract with the Water Environment Research Foundation (WERF) explored and analyzed public acceptance and public involvement in the field of biosolids recycling in North America. For this report, the term "biosolids" is used to mean municipal sewage sludge solids (sludge) that have been treated and tested and are being used as fertilizers and/or soil amendments. Four tasks were completed before the writing of a final analysis and recommendations.

1) Investigators collected about 1,000 references from peer-reviewed, published and gray literature regarding public perceptions and interactions with biosolids recycling programs and social science literature on communications and public perceptions and participation. A searchable electronic, annotated bibliography is **included on the enclosed CD-ROM.**

2) Investigators developed 14 case studies that document and analyze a variety of experiences with public perceptions and public involvement in biosolids recycling programs. The details of the case studies are **included on the enclosed CD-ROM.**

3) Based on the literature review and case studies, the investigators developed the first-ever, large-scale national survey regarding public perceptions of biosolids recycling. The University of New Hampshire Survey Center conducted the survey by telephone in early 2002, collecting opinions from more than a thousand U.S. residents. The details of this survey are **included on the enclosed CD-ROM.**

4) Investigators held a one-day workshop the Boston area in April 2002 in order to bring additional expertise from the social sciences to the project. At this workshop, the project team and advisors from around the continent learned more about how public perception develops, what affects the public's understanding of and response to a proposed project, and how biosolids managers might better interact with the public to gain greater trust and improve their programs. Details from this workshop are **included on the enclosed CD-ROM.**

The results of these four steps were brought together into this document, Developing Public Participation and Earning Trust.

What follows are summaries of the findings from each part of the project, which are detailed further in the subsequent chapters of this report and can be found on the enclosed CD-ROM.

Literature Review

The literature review was conducted to learn what has been written about biosolids recycling by those in the biosolids management field as well as interested persons outside the field. The scope of the literature search was narrowed to documents published or presented mostly after 1990 and mostly in the United States and Canada. The included literature pertained to 1) public knowledge, perceptions, opinions, and behavior about biosolids recycling, and 2) the most significant and summary literature regarding public knowledge, perceptions, opinions, and behavior around related environmental issues (e.g., siting of facilities, solid waste management, energy, and new technologies). A wide variety of perspectives on environmental issues and biosolids recycling were included.

History of Public Acceptance

Beginning in the 1970s, the U.S. EPA, state agencies, and managers of municipal sewage sludge began to recognize the importance of public perception and public acceptance. Technical manuals and guidance documents began to include short sections on working with the public regarding the use or disposal of sewage sludge.

By the early 1990s, "public acceptance" had become a regular topic of interest in the trade press and at conferences. As public interest in civic and environmental affairs was generally increasing around North America, a growing number of conflicts between biosolids recycling programs and local communities were underscoring the importance of public acceptance. "A technically well designed and executed program can still fail because of community opposition," noted one author.

The Water Environment Federation (WEF) tackled the public acceptance challenge through its publications, including a biosolids communication plan developed by a Washington public relations firm (Powell-Tate, 1993). The "Powell-Tate" communications plan encouraged biosolids managers to identify and communicate proactively with "gatekeepers," individuals who are leaders of public thinking in their communities (political leaders, agricultural advisors, conservationists, civic club leaders, etc). Efforts to gain public acceptance stressed the importance of conveying the abundant scientific information on which biosolids recycling practices were based, especially with the help of allied gatekeepers. The assumption was that if people learn more about biosolids recycling, they will come to accept it.

Pushed in part by public concerns, U.S. EPA engaged the National Academy of Sciences (NAS) to conduct a review of the science behind the federal 40 CFR Part 503 regulations—the 1993 regulations that govern sewage sludge management in the U.S. That 1996 report confirmed what proponents had been saying, that, according to the best available science of the time, biosolids recycling involved only "minimal" risk. But that scientific information did not quell the conflicts, and the NAS panel knew it would not. They recommended that biosolids recycling program managers pay attention to "early public involvement in the design of land application programs"; support of strict enforcement, oversight, and best management practices*;

---

* The term "best management practices," as used in this report, refers to a variety of actions that those managing and using biosolids take to help ensure that any potential risks or negative impacts from their biosolids recycling program(s) are reduced as much as possible. Best management practices ("BMPs") develop over time, through general agreement within the field, and are usually based on research findings. Occasionally, researchers, regulators, or agricultural advisors (e.g. cooperative extension) compile into a guidance document current best management practices for a particular aspect of biosolids recycling (e.g. U.S. EPA's *Guide to Field Storage of Biosolids*). BMPs

demonstration programs; indemnity of farmers; and empowerment of local and state officials to address and enforce nuisance and other complaints.

Soon after the publication of the NAS report, the Cornell Waste Management Institute circulated a paper called "The Case for Caution," a critique of biosolids recycling that raised many questions about its safety and long-term impacts. Local newspapers, which had covered biosolids recycling controversies in scattered locations, began to be joined by occasional regional and national press coverage. By the turn of the millennium, failure to gain public acceptance was becoming a relatively common reason for the failure of biosolids recycling programs. Local and county restrictions or bans were being enacted from New Hampshire to southern California. In response to the growing public concern, regulations were strengthened in many states.

In response, the biosolids "industry" formed a few regional biosolids management associations and implemented training programs and developed a progressive Environmental Management System for biosolids program to help reduce nuisance problems. Communications to the public were stepped up—more videos, fact sheets, and brochures appeared. U.S. EPA requested another NAS review; it was completed in July, 2002.

As opponents to biosolids recycling became more organized and effective through publications of their own, including use of Internet websites and email, more biosolids managers and regulators began to recognize that biosolids recycling nationwide was strongly threatened by the "public veto"—the ability of small groups of committed citizens to put social and political pressure on local and regional citizens, regulators, and legislators to stop particular projects. Discussion in trade journals and conferences turned to the need for increased public participation in developing and implementing biosolids recycling programs.

Topics in the Biosolids Recycling Public Acceptance Literature

Much of the literature on public acceptance of biosolids recycling fell into the following general categories:

- *Scientific and public health issues.* These include the potential impacts (and how to discuss them) of trace elements ("heavy metals"), chemicals, and pathogens in biosolids; odors and air quality; oversight and enforcement; surface water and groundwater quality; soil and food quality; transportation and trucking; economic viability; changes in demographics and changing expectations; and emerging issues and uncertainty.
- *Examples of biosolids recycling programs that gained public acceptance.* Biosolids management professionals have written many "success stories" about biosolids recycling programs. These are widely used to promote the benefits of the practice.
- *Biosolids recycling program communications and outreach. How much and what type of biosolids information does the public need? Who should speak for biosolids recycling?* Building on the Powell-Tate communications plan and other guidance documents, an increasing number of conference and trade press papers urge biosolids managers to communicate better and provide better information. Some discuss *how* biosolids managers may best communicate the biosolids recycling "story."
- *Political and legal efforts.* A few papers in the literature describe political and legal efforts against or in defense of biosolids recycling.

---

develop and improve over time. High quality biosolids management is knowledgeable about—and utilizes—the most current BMPs available.

- *Public opinion surveys and biosolids marketing analyses.* Several biosolids recycling programs in North America have conducted surveys or marketing analyses to better understand public perceptions and concerns around biosolids products. However, only two small *nationwide* surveys were found in the literature–one conducted as part of the development of the WEF communications plan by Powell-Tate and the other as a later check on progress from that plan.
- *Addressing issues through Environmental Management Systems (EMS).* As the National Biosolids Partnership began to develop its EMS program in the late 1990s, with the goal of encouraging "practices that build public confidence within local communities," papers began to discuss the potential benefits of an EMS in terms of public acceptance.
- *The roles of third parties and the media.* As the media took more notice of biosolids recycling—especially conflicts around it—biosolids managers wrote more about "how to deal with the media."
- *Literature expressing concerns about biosolids recycling.* Any review of the literature on public acceptance of biosolids recycling must include the extensive and growing number of articles, papers, fact sheets, and Internet sites that express concerns about the practice. Besides "The Case for Caution," a book on public relations called *Toxic Sludge is Good For You* became popular amongst critics. Many small civic and environmental groups have printed reports critical of specific projects or biosolids recycling in general. An Internet search using the keyword "sludge" will access a wide variety of information, much of which pertains to biosolids recycling, but some of which is about other kinds of sludge (dredged sludge, oil sludge, hazardous waste sludge, etc.). This literature expressing concern has grown in volume and impact over the years. How to counter it has become an on-going public acceptance topic in the biosolids management field and literature.

Social Science Literature

The project team found only a handful of social science studies that had been conducted to look specifically at the social factors involved in public perception and acceptance of biosolids recycling. However, the literature review includes social science research topics that are relevant to biosolids recycling, including:

- how people perceive potential risks ("risk perception"),
- risk communications,
- trust,
- equity and fairness,
- public participation,
- conflict and conflict resolution, and
- general public attitudes regarding the environment and technology.

The modern literature on these and related topics dates back only as far as the late 1970s. Since then, social scientists have increasingly sought to understand behavior change and the relationships between the public's knowledge of a subject and their attitudes, perceptions, and behavior—especially in the environmental fields of energy, nuclear power, and hazardous waste management. What they found was that public perceptions about energy and the environment were not shaped exclusively by technical or scientific information. They also noted that, over time, the public was becoming more skeptical of new technologies and their perceptions were not the same as those of the "experts" (i.e., scientists, engineers, and government officials).

Social scientists Peter Sandman, Paul Slovic, and others researched risk perception and developed the concept of "outrage" factors that influence how an individual perceives a particular risk (risk = hazard + outrage). How best to communicate risks was an area of study that developed from this understanding of risk perception. Others, such as Lawrence Susskind, looked at conflicts around environmental issues and examine individual behaviors in terms of trust, credibility, equity, fairness, and respect. These were identified as key factors in developing productive interactions that led to resolution of conflict or avoidance of conflict. Susskind and his associates developed and have written extensively about the "mutual gains approach" to resolving conflict, which emphasized trustworthy actions and a focus on building long-term relationships.

Through the 1990s, such concepts were developed further and reached into specific fields—for example, the health field and journalism has utilized risk communications extensively.

At the same time, the public has come to expect, more than ever, to be able to be involved in public sector decisions, if they choose to be. Thus, "public participation" has become a major topic in diverse environmental and public policy fields. There have been many different attempts and models for how to involve the public most effectively. The International Association for Public Participation (IAP2) has attempted to standardize strategies and tools for use in fostering public participation in any field. Most recently, social scientists have been evaluating the effectiveness of different public participation models. One researcher (Rowe and Frewer, 2000) recommends the application in the public participation process of certain criteria that he believes make a model more acceptable to the wider public:

- representativeness (how well does the participation effort manage to involve all representative stakeholders?),
- independence (is the public participation group able to act and make decisions independently of any one stakeholder group?),
- early involvement (people tend not to want to be involved if they cannot participate early in the process, when key decisions often are being made),
- influence (is the public able to have real influence on the outcome?), and
- transparency (is the process by which decisions are made and actions taken transparent to all involved?),

He also recommends process criteria that he believes make a model more likely to be effective:

- resource accessibility (do all participants have fair and equal access as needed to information and other resources?),
- task definition (is the task of the public participation group or effort well defined and well understood?),
- structured decision making (is their a clear, defined, agreed-upon process by which decisions are made?), and
- cost-effectiveness (will the public participation process avoid breaking the bank?).

The complete literature review appears on the enclosed CD-ROM, in addition to the annotated bibiliography.

Case Studies

In consultation with a group of biosolids recycling stakeholders convened for this project, the investigating team chose to develop case studies about public acceptance issues and efforts at fourteen different biosolids recycling programs around North America. The experiences described in these case studies range from programs that have developed considerable public support by developing effective communications, trust, and long-term relationships with key "gatekeepers" (e.g., King County, WA) to programs that were stopped by public outrage (e.g., Kern County, CA). The case studies provided additional insights into how public acceptance issues noted in the literature review play out as biosolids recycling programs interact with the public (Table I-1).

Table I-1. Public Acceptance Issues Identified in the Biosolids Case Studies.

Public Acceptance Issues	1. King Co., WA	2. Milwaukee, WI	3. Jefferson Co., AL	4. Princeton, BC	5. Fulton Co., IL	6. Boulder, CO	7. New York City CAC	8. NYC and Rural Communities	9. Denver,CO	10. Philadelphia, PA	11. Montgomery Co., MD	12. Everett, WA	13. Kern Co., CA	14. New Hampshire
Social/Political														
Trust	•		•	•		•	•	•	•			•		•
Equity and Fairness	•		•	•	•	•		•	•	•	•	•	•	
Communications	•	•	•	•				•	•	•		•		•
Information (Sources and Timing)	•			•			•		•			•	•	•
Public Participation	•			•	•	•	•	•	•		•	•	•	•
Changing Demographics	•		•			•			•	•	•			•
Politics									•		•		•	•
Oversight and Enforcement	•							•	•				•	•
Scientific/Public Health														
Trace Metals and Chemicals	•	•						•	•	•		•	•	•
Pathogens								•		•	•	•	•	•
Odors and Other Air Quality				•	•		•			•				•
Water Quality			•	•					•	•		•		•
Soil and Food Quality										•			•	
Transportation/Trucking			•	•				•	•				•	•
Economic Viability			•		•		•	•	•		•	•	•	•
Emerging Issues and Uncertainty													•	•

The highlights of each case study are as follows:

1. *King County, WA*—The Seattle area's long-standing biosolids recycling program has relied on strong partnerships with allies: university researchers, cooperative extension staff, local community leaders, farmers and farming groups, and a conservation consortium—the Mountains-to-Sound Greenway Trust. The biosolids program has high standards for the management of its biosolids, applying best management practices and conservation measures to minimize any potential negative impacts, including nuisances. Applications of biosolids on forestry and agricultural sites are developed and researched to show clear benefits. King County has invested millions of dollar in research and public outreach. Staff have spent much time and effort introducing biosolids projects to local communities. Regular, multi-pronged communications keep stakeholders informed. Questions and concerns are addressed promptly, through the Northwest Biosolids Management Association, university research and demonstration, and/or by knowledgeable County staff. The program's allies are willing to speak in support of the program. King County's biosolids program consciously implemented the Powell-Tate (1993) communications plan and staff believe that has led to the stable long-term biosolids recycling projects the County has today.

2. *Milwaukee, WI*—Milorganite®, a bagged, pelletized, heat-dried biosolids fertilizer product, has been marketed since 1926. Its success has come from its performance as a pleasing, easily-managed, effective commercial product—it stimulates plant growth and customers are happy with it. This level of customer satisfaction has come about because of Milwaukee's recognition, decades ago, that they create a *product* that must meet the demands of the consumer. Thus, Milwaukee's biosolids program has a consumer-oriented organizational ethos that understands and responds to its customers, requiring proactive communication, listening, and constant attention to quality and improvement.

3. *Biosolids Mine Reclamation in Jefferson County, AL*—This county biosolids recycling program, serving Birmingham and neighboring communities, established a land reclamation project in Beltona that provided clear benefits to the land and neighbors. However, when the County proposed the same kind of project at another site in the town of Adger, neighbors objected to the city "dumping on them" and worried that their wells or children might be harmed. "It was a natural reaction to something they didn't know about," said biosolids program staff. This case study indicates that, with concerns about biosolids being openly and frequently expressed on the Internet and in other media sources, past successes are no longer predictors of future public acceptance. On-going outreach and public involvement is required to ensure the long-term viability of any biosolids recycling program.

4. *Princeton, BC*—Clear benefits to the local community was the hallmark of the Greater Vancouver Regional District's reclamation with biosolids of the Granby mine tailings at Princeton, BC. The project managers emphasized this perspective as they communicated extensively with the local media and residents, addressing questions and concerns openly and honestly. The project started with a small demonstration, to show the potential benefits of the project, and then grew to become a project noted by the British Columbia Lung Association. In an article in that association's publication, *Your Health,* they stated that the biosolids application had significantly reduced airborne dust and some people have now moved into the area for the air quality.

5. *Chicago Biosolids at Fulton County, IL* —In the early 1970s, the Metropolitan Water Reclamation District of Chicago purchased more than 15,000 acres of mine-spoiled land in Fulton County, south of the city. A local steering committee was formed to assist the District in developing the project to reclaim and restore the mine-lands with biosolids. District staff who managed the operation lived near the site. Early operations created odors and public concerns, which took many improvements to practices and years of operations to overcome. Today, even though the Prairie Plan biosolids recycling program is one of the longest continual land application programs, and has yielded clear benefits to the land, local public acceptance is limited. "Local people are still not thrilled about it coming from Chicago," says an Illinois regulator.

6. *The Study Review Group, Boulder, CO*—When Boulder's biosolids land application program met strong local resistance in 1994, the staff convened a "Study Review Group" of diverse stakeholders that was given the task of finding an acceptable method of biosolids management. The group was provided independent technical consultation and a neutral facilitator. They met regularly for months, learning about various options, and eventually agreed on an interim plan that was presented to the City council for approval. The interim plan included a combination of Class A and Class B biosolids recycling, with the proportion of Class A composting to be increased once a pilot compost operation was completed. The outcome and the process were considered fair and productive by all involved, and controversy around the topic subsided after the interim plan was approved and began to be implemented. In the end, the City was able to adhere to its policy goal of recycling 100% of the sewage sludge it produced.

7. *New York's Citizen Advisory Committee*—In 1989, New York City established a diverse Citizen's Advisory Committee (CAC) for sewage sludge management that included representatives of environmental groups, community and borough board members, business representatives, city staff, and more. Committee members brought different interests and agendas to the process, but all recognized their responsibility for developing an agreed-upon Biosolids Management Plan. The City provided the CAC independence and technical support. The CAC commented on, and revised, the Plan, which eventually was approved and implemented. Thereafter, when issues arose, the CAC responded, defending the biosolids management program. A revised CAC still meets six times a year to review new information and make additional recommendations. Researchers have noted that the effectiveness of CACs is improved by agreed-upon ground rules, wide and equitable representation, transparency of the process, early involvement in the decision-making process, opportunities for full learning of technical details by all CAC participants, and full sharing of information.

8. *New York City and the Rural Communities of Sierra Blanca, TX, and Prowers County, CO*—Part of New York City's biosolids management plan has been to transport biosolids long distances for agricultural use in rural areas in western states. Biosolids managers recognized that overcoming concerns about contaminants in the City's biosolids would be a significant obstacle. They took actions to overcome this: stepped-up pretreatment, reducing the corrosiveness of the City's drinking water, strict monitoring of the biosolids, and garnering support from their Citizens Advisory Committee. In addition, City staff insisted that they be visible and involved in all communications and interactions with the communities where the biosolids were going to be recycled. For their part, the biosolids management companies that conducted the land application

programs worked with local community members to develop practices that would help allay local concerns, including a container inspection and handling plan to ensure that nothing but actual biosolids could be slipped into shipments. Extensive communications were conducted. Local farmers, ranchers, and officials were well-informed, so that some became proponents of the biosolids recycling programs. Another critical component of the success of the Prowers County program was that the host community experienced direct financial benefits through decreased fertilizer costs and improved soils and crop yields. These two New York City biosolids projects highlight critical factors that lead to gaining public support:

- New York City worked with contractors who had the resources to do the job;
- City staff have been partners in the program, assisting the contractors in maintaining credibility;
- The contractors identified and understood the relevant local issues and socio-political situations and addressed them appropriately; and
- Both contractors utilized local resources to address local issues.

9. *Denver, CO's Metrogro Farm Biosolids Land Application Program*—Understanding and addressing the needs of the local community was a central element in making Denver's Metrogro Farm in Deer Trail, CO, a viable site for on-going land application of Class B Denver biosolids. When Metro bought their own huge tract of land for their biosolids recycling program, they expected to have little local public resistance. But neighbors and civic leaders became concerned about their farming methods. When an order to stop land application of biosolids came from the County Supervisors, Metro decided not to bring legal action, but, rather, to try to better understand and address the concerns. Over several years of dialogue, many tours of the site, and the development of an extensive independent monitoring program by the U. S. Geological Survey, relations between Metro and the local community of Deer Trail have improved. Lessons learned include:

- Actions speak louder than words—the most important thing a biosolids program can do is to constantly strive to do the right thing, to pay attention to every detail ("doing what we say we will do—not just lip service!");
- Listening is more important than talking--it is critical to understand that critics are often right and can provide valuable feedback;
- There is no easy, canned solution to gaining public support.

10. *The Philadelphia Biosolids Mine Reclamation Program*—Philadelphia biosolids have been used for more than 30 years to restore barren lands scarred by past mining. This use has included extensive study by university researchers. Yet considerable public concern has developed throughout the state, and opponents have become sophisticated and effective:

- Concerned citizens have organized around the issue;
- They point to the origins of biosolids and the uncertainties involved in recycling them;
- They encourage distrust of those producing and managing biosolids;
- They have convinced some larger regional organizations and state legislators to oppose biosolids recycling;
- The Internet has made communication amongst those concerned much easier;
- Opponents point to two deaths they believe were caused by land application of biosolids; and
- They draw support from some experts in the field, notably EPA "whistleblowers."

As a result, maintenance of the Philadelphia mine reclamation program has become far more complex than it once was. Contractors must now pay constant close attention to local and regional social and political factors, more open and regular communications with more different stakeholders, addressing questions and concerns, and understanding the interests of concerned citizens.

11. *The Washington Suburban Sanitary Commission Site 2 Montgomery County Regional Composting Facility*—Siting of this composting facility in the small bedroom community of Calverton, MD (outside of Washington, D.C.) was fraught from the start with controversy and legal battles. In 1982, the site finally began receiving and composting sewage sludge; odors were an immediate problem. Over the next 10 years, a variety of expensive odor control measures were added, amidst mounting local public anger and pressure. In 1999, local opposition had created enough political power to force closure of the facility. Several lessons can be learned from this case:

- Proper siting of any facility is extremely important.
- Communities change, and a site that was once good may no longer be.
- A new technology and/or facility should be scaled up in phases in order to evaluate and deal with potential problems.
- Citizens should be engaged in the development process as early as possible.
- Involving the public in planning and decision-making can often result in better, more efficient, and cost-effective decisions.

The experiences gained at Site 2, the nation's first large-scale municipal sewage sludge composting facility, have assisted in establishing better management practices elsewhere.

12. *Everett, WA*—Biosolids managers from the City of Everett have reflected on their failure to successfully develop a forestry biosolids land application program on the Tulalip Indian Reservation. Although the City staff notified and involved local community members early in the development process, they failed to involve some key community leaders. More importantly, they were not well trained and were unable to answer questions and concerns in a timely manner. In addition, some social and political factors were overlooked; for example, the agreement worked out between the City and the tribe turned out to be politically and socially unacceptable to many people in the area. Finally, City staff did a poor job of gauging the mayor's and city council's commitment to the project; when the public outcry grew, their support waned. Key lessons from this case study include:

- Being open and honest was not good enough; there needed to be greater organizational commitment of resources (e.g., staff time and training);
- A successful project requires staff to cultivate strong support from elected officials;
- Local sponsors for the project must be treated as partners;
- Community leaders must be contacted early;
- These and other opinion leaders must be kept informed and their support nurtured;
- City staff must be very knowledgeable before the project is announced;
- The public must trust that the City knows what it is doing;
- Information must be timely, and questions about the project must be answered promptly.

13. *Kern County, CA*—In late 1999, Kern County, an agricultural county in south-central California, adopted an ordinance banning land application of Class B biosolids. The ban

threatened major biosolids recycling programs of Los Angeles and Orange County. Issues that led to the ban included:

- The lack of any education and outreach campaign, which resulted in entrenched, unchangeable opposition exemplified by the comments of a Kern County farmer who stated, "We don't want to be L.A.'s toilet;"
- The impossibility of proving zero risk;
- Apparent scientific uncertainty regarding the safety of biosolids recycling;
- The opposition of local carrot growers because of their fear of public perception associating biosolids with their produce;
- Use of low-bid contractors who took shortcuts to reduce costs, which led to local environmental concerns; and
- Lack of monitoring of contractors and practices.
- The biosolids program contractors failed to recognize significant political and social motivations in Kern County and how to deal with them. Since the conflicts of 1998 and 1999, it has taken several years to begin to reestablish some trust amongst stakeholders.

14. *How Biosolids Recycling Failed and Rebounded in New Hampshire*—When the federal Part 503 regulations were adopted in 1993, New Hampshire chose not to regulate biosolids beyond the federal rules. Because Part 503 has few requirements for managing biosolids in ways that do not cause public nuisance, New Hampshire quickly experienced conflict between some biosolids managers who failed to avoid the generation of malodors and other nuisances and communities in which land application programs were occurring. In addition, the following factors were in place and helped create public concern:

- Lack of oversight and enforcement at the national level due to the low priority placed by EPA on oversight of sewage sludge use or disposal.
- A relatively young biosolids management market and infrastructure, involving many competitive private firms and numerous wastewater treatment facilities trying to find the most cost-efficient ways to manage sewage sludge, creating incentives for cutting corners and trying untested management processes.
- A prevailing concept within the wastewater management field that sewage sludge was something to be disposed of in the most cost-efficient manner possible.
- A public mostly unaware of wastewater treatment and sewage sludge management.
- A rural state feeling the social effects of growth, with development impinging on agricultural lands, generating local conflicts.
- A strong tradition of local control and skepticism (even distrust) regarding outsiders and government.

By 1995, conflicts were occurring around the state. Biosolids recycling became a political issue, even in the 1996 and 2000 gubernatorial campaigns. Opponents convinced more than 30 towns to impose significant restrictions or outright bans on biosolids recycling. This spurred biosolids managers, wastewater operators, and state regulatory officials to work harder at addressing the conflict through meetings and advisory groups. Although it was a difficult and frustrating public debate for all involved, increased mutual understanding occurred. Once new rules were adopted in 1999, fewer complaints ensued and conflicts diminished.

Today, New Hampshire biosolids managers better understand the perspective of concerned environmentalists–concerns about the long-term impacts of trace metals and chemicals applied to the land in biosolids products, the potential impacts on public health of trace pathogens (especially in Class B biosolids), and concerns about the amount of testing, oversight, and enforcement being done to ensure that existing regulations are being followed. For their part, many critics of New Hampshire's biosolids recycling programs have come to a better understanding of the limited options for cost-effective sewage sludge management and the challenges faced by water quality professionals working with wastewater and biosolids on a day-to-day basis. The mutual respect that has developed amongst some people on various sides of the issue is a sign of the development of important relationships that could lead to more productive and mutually satisfying progress on sewage sludge management issues in the future.

Details on these case studies can be found on the enclosed CD-ROM.

The 2002 Biosolids Knowledge and Perception Survey

Based on the understanding developed from studying the literature and case studies, the investigating team developed a nationwide survey questionnaire, with assistance from the University of New Hampshire Survey Center. The Survey Center conducted the formal survey by telephone utilizing a nationwide sample that, in the end, totaled 1069 individuals. Those surveyed were selected at random from the nationwide pool of home owners and house renters—people thought more likely to know something about sewage treatment and agriculture/gardening. The phone interviews were conducted in February, 2002. The results of the survey were developed by the Survey Center using standard statistical methods. Significant findings include:

- Knowledge of the term “biosolids” is limited (only 3% can accurately define "biosolids" and another 11% had a fairly good idea of what “biosolids” are).
- Support for wastewater treatment is very high (93%).
- Regional differences are minimal regarding the level of public knowledge and perceptions of biosolids recycling.
- People are uneasy and have questions about biosolids recycling.
- The choice of words makes a significant difference: using "sewage sludge" instead of "biosolids" creates a marked drop in positive response to a neighbor's hypothetical use of the material.
- If they need to learn more about biosolids, people want more information about many different aspects of biosolids recycling.
- Faced with a hypothetical situation in which biosolids are used by a neighbor, people say they would turn to and trust friends and neighbors, government agencies, and academic researchers for initial information.
- Faced with a hypothetical situation in which biosolids are used by a neighbor, people say they would trust government and others who appear knowledgeable and objective; they strongly distrust those with a profit motive.
- People favor constructive uses of biosolids (creating energy, recycling nutrients).
- Some factors increase people’s concern about biosolids (if it includes an industrial waste source or is from a large city), others decrease their concern (if it is certified annually, if they are contacted prior to its use, if it is supervised locally).
- Strong arguments in favor of biosolids recycling are: Biosolids recycling returns nutrients to the soil, and Recycling biosolids disposes of a necessary waste.

- Most people think that biosolids recycling is a relatively small risk to them.
- Although they clearly expressed support for the concept of recycling biosolids, respondents indicated that the strongest argument against biosolids recycling is "not enough is known," indicating that they are uncomfortable with their own lack of information and/or what they perceive to be an overall lack of information on this topic.

The 2002 Biosolids Public Knowledge and Perception Survey provides biosolids managers and other stakeholders with information about many factors and actions that will impact their chances of building trust, improving communications, and building public support for biosolids recycling programs. There is no silver bullet—no one thing that will make the difference in terms of improving public perceptions, participation, and support for a particular biosolids management program. Rather, there are many small actions and factors that a biosolids manager can pay attention to—and the total sum of those actions will increase the likelihood of gaining greater public participation, greater public trust, and improved public perceptions of any particular biosolids management program.

The CD-ROM contains the summary and interpretation of the data from the survey. The survey questionnaire and the compiled raw survey data are included in Appendices B and C, respectively.

Workshops on Public Participation

The reviews of literature and case studies and the results of the survey pointed the investigators to the need for greater interaction with the public on the part of biosolids management programs. How to accomplish that became the final important research goal of this project.

One important aspect of developing stronger public input and participation in biosolids management is training. How to interact with the public is not a common training topic in the wastewater management and engineering fields. Therefore, the project sponsored a one-day workshop to provide initial training on conflict resolution and the "mutual gains approach." In attendance were project advisors, biosolids managers, and other stakeholders from around North America. The summary from that workshop appears in Appendix D. Lessons learned from this and other workshops conducted subsequently at Water Environment Association conferences have been incorporated into the concepts and recommendations included in this report.

Conclusion: Developing Public Participation and Earning Trust

This report provides recommendations for how biosolids managers can increase constructive relationships with communities, neighbors, and other public stakeholders. Over the course of this project, the investigators changed their perspective from a focus on finding ways for biosolids managers to gain "public acceptance" of their biosolids recycling programs to a focus on "developing public participation and earning trust"; the recommendations reflect this change in perspective. The two approaches involve significantly different assumptions, speech, and actions.

> **The "public *acceptance*" approach**—One-way communication: "I am the expert. I know about biosolids. I will give you more information and educate you and then you will agree with me."
>
> **The "public *relationships*" approach**—Two-way communication: "Let me hear your concerns…. O.K., I think I understand. Can we discuss this more? I want to help address your concerns. May I share with you what I know and suggest who else you might talk with? How can we start?"

The ultimate goal of a biosolids recycling program's outreach, education, public relations, and public participation efforts should be to develop "public relationships" with the variety of stakeholders who may be involved with and/or potentially impacted by a biosolids recycling program. Developing relationships through listening to, and having dialogue with, all stakeholders leads to increased trust and respect, as well as increased public understanding of biosolids recycling. In most instances, this relationship-building approach also will build better biosolids management programs that are supported widely by a diverse public.

What is the Problem?

The conflicts that develop around biosolids recycling are uncomfortable for everyone. People become emotional. People get angry. Biosolids managers feel defensive, hurt, indignant, right, wrong, uncertain. For many, this is their first experience with public conflict, and it doesn't feel good.

From the public's point of view, the recycling of biosolids is a complex concept that many people find difficult to understand and accept. It is rife with factors that create uncertainty—what exactly is in any one truckload of biosolids? Because no one knows all of what gets put down drains into sewers, could there be harmful quantities of organic chemicals or "heavy metals" in some biosolids? What will be the lasting effects of adding cumulative amounts of "heavy metals" to soils over many years? What about the pathogens in biosolids?… And uncertainty is just one of many things about biosolids that can raise the level of some people's concern.

Further, while much of the wastewater treatment process requires little interaction with the general public, biosolids recycling is different—it extends wastewater operators' and engineers' work right into communities and the consumer marketplace. If biosolids managers are not aware of the affects of real and perceived uncertainty, they have trouble responding constructively to natural human concerns when they are forced into interactions with concerned citizens.

Too often, biosolids managers feel that the public's questions and concerns are threats to their biosolids recycling program and their efforts to protect water quality. They react defensively.

Instead, biosolids recycling program managers need to look at these and other "social" factors that affect a biosolids recycling program. If they don't, they will likely end up, sooner or later, getting some surprising and unpleasant feedback from the public. Sometimes, as seen in some of the case studies, that feedback is so strong that biosolids programs—even ones that have strong technical merit—are stopped. This "public veto" of biosolids recycling programs has happened all too often—and mostly because of the failure of biosolids managers to recognize the importance of public interests and concerns.

What's a Biosolids Manager to Do? Overarching Recommendations:

Managing sewage sludge is an important public function in modern society. Sewage sludge and biosolids managers should recognize that:

- they do not have to decide all by themselves how to manage sewage sludge and/or biosolids;
- if they want to ensure that the decisions being made are the best possible—and the most publicly supportable—they should get input from a variety of stakeholders (i.e., they should strive for and insist on public participation);
- if they are going to involve the public in meaningful, constructive ways, they must build "public relationships" with individual stakeholders; and
- building public relationships requires attention to the relationship-building factors of trust, fairness, honest and useful information, two-way communications, and organizational commitment and motivation to do things right for most stakeholders.

In short, all public outreach, communications, and participation efforts and actions, as well as many technical decisions regarding biosolids management, should be developed and implemented with this question in mind:

"What will be the impact of my next action on our program's relationship(s) with key stakeholders?"

If the impact(s) will be negative, consider whether you should continue with that action. How critical are the relationships that will be negatively impacted? Remember that, if trust is broken, it can take years to regain. Are there other actions you can take—e.g., talking to the impacted stakeholders in advance—that will help lessen the negative impact on the relationships? Sometimes you have to take actions that will be unpopular with some people—but it is best to do so knowing what the reaction will likely be. And, if possible, it is always best to let people know in advance.

Today, for many biosolids programs, the first decision has to be to come out from under the radar. Biosolids recycling programs are too often hidden. Biosolids managers often feel that "as long as no one is noticing and no one is complaining, all is well." Such programs are vulnerable, at any moment, to discovery and sudden public outcry.

It is hard to do, but in the long run, making your program more visible, speaking openly of its benefits and risks, and sharing information widely with diverse stakeholders will help to ensure its long-term stability. The benefits to biosolids management programs of improving public relationships are:

- you will have others willing to stand up and support your decisions because they were involved in those decisions;
- you will increase the likelihood of avoiding mistakes;
- you will reduce the vulnerability of your program to criticism;
- you will find more efficient and environmentally beneficial solutions through the diverse input you receive.

The report discusses 12 steps for developing and implementing a public outreach and participation program (Table I-2). Many of the recommendations are not new, but their consistent implementation by biosolids managers has not yet been realized.

Table I-2. Twelve Steps for Developing Public Participation and Earning Trust.

Step 1.	Build commitment within your organization.
Step 2.	Understand the national and local context.
Step 3.	Learn more about conflict.
Step 4.	Determine what is beyond your control; work on what you can control.
Step 5.	Determine who your stakeholders are.
Step 6.	Get input from key stakeholders as soon as possible, early in the process.
Step 7.	Redefine success.
Step 8.	Build public relationships.
Step 9.	Improve communications, especially listening and dialogue.
Step 10.	Provide information useful to *recipients*.
Step 11.	Don't be afraid of conflict—work patiently through it.
Step 12.	Continue to monitor public perceptions and public relationships.

Key Principles to Guide Public Participation

Biosolids managers are encouraged to pay attention to social aspects of their programs. An overarching guiding principle is to say things and take actions that build constructive long-term public relationships. Building positive public relationships hinges on the following principles, which have become central WERF themes identified from social science literature and experience through this project and a related one. Think of public relationships as an umbrella held up by:

- trust;
- fairness;
- quality of information;
- excellent communications; and

- your organizational commitment to public outreach, dialogue, and participation.

- In addition, the following factors are important for biosolids managers to keep in mind.:

- ***Listen*** to stakeholders; encourage their input; never discount or scoff at questions or concerns—most questions are reasonable to those raising them.
- ***Demonstrate that you care*** about protecting the environment. Ensure continual improvement in practices, and consider how you can remove factors that contribute to nuisances or public outrage (e.g., size and site projects appropriately). An Environmental Management System can help institutionalize continual improvement.
- ***Ensure independent oversight***, either by a state or county agency or by creating independent oversight at the local level: it builds public trust, credibility, and confidence in your program.
- ***Give the public time to learn***; big trouble comes when people feel pressured by looming deadlines. This may mean communicating extensively and getting public input long before any required public hearing or application deadlines, so that people are already well-informed when the deadlines arrive.
- ***Engage stakeholders as early in the process as possible***: since biosolids recycling is a public function, interested citizens have a right and a responsibility to constructively assist the biosolids manager. Engaging constructive stakeholders may be one of your biggest challenges—so start working on it as soon as possible.
- ***Focus on building long-term relationships***: Remember to ask yourself "What will be the impact of my next action on our program's relationship(s) with key stakeholders?"
- ***Remember, you are not giving up responsibility or control*** by reaching out and involving stakeholders. You are still ultimately responsible, but you will have greater confidence (and your program will be less vulnerable) knowing that different stakeholders know about, have been involved in, and accept what you are doing.

CHAPTER 1.0

DEVELOPING PUBLIC PARTICIPATION AND EARNING TRUST

Ned Beecher
Nora Goldstein

"People wake up to this going on in their neighborhood, and no one from your industry has gone to them to explain what it is. And they become afraid. People are going to be concerned.... People have a right to be concerned. And, what I get when I ask questions is a barrage of research papers and lots of facts, but what's left is still that people are concerned, and nobody is talking to them."
--Sally Garner, Producer, CBS Evening News

1.1 Introduction

It has been more than 30 years since the passage of the Clean Water Act and more than 10 years since the promulgation of 40 CFR Part 503, the federal Environmental Protection Agency (U.S. EPA) regulations for the management of municipal sewage sludges. The use of biosolids—treated sewage sludge in applications to soils as fertilizer and soil amendments has become common across the U.S. and Canada. Today, nearly 60% of U.S. sewage sludges are applied to soils in biosolids recycling programs (National Research Council, 2002). Yet relatively few people are aware of biosolids recycling programs. And those who are aware have varied perceptions of biosolids recycling. Some have expressed considerable concerns, and organized opposition is challenging the viability of a number of biosolids recycling programs.

The research leading to this report has provided an overview of the current state of public perceptions and public acceptance of biosolids recycling in North America.See the accompanying CD-ROM for the detailed literature review (Chapter 2.0), case studies (Chapter 3.0), comprehensive national telephone survey (Chapter 4.0), and associated appendices The results of that research confirm that biosolids recycling programs that find strong support from a variety of stakeholders[1] tend to have the following characteristics:

[1] The term "stakeholder" is used in this chapter to refer to those most directly involved or potentially impacted by a biosolids management program, including farmers using biosolids, agricultural advisors, immediate neighbors, etc.

- they are compliant, well-run operations following best management practices[2];
- their benefits to communities and the environment are obvious;
- they introduce themselves to stakeholders early in the siting process and employ a respectful, joint-learning process, often led, in part by local farmers or other citizens;
- they communicate well through a variety of methods with the variety of stakeholders;
- they provide consistent, high-quality, receiver-appropriate, timely technical information to anyone interested;
- some incorporate significant public participation (or customer feedback) in program planning; and
- they have strong organizational commitment (staff time and funding) behind their public outreach efforts (including from the public agency producing the biosolids).

Other programs, however, have failed to work well with and for the public, and conflict has sometimes become intense and debilitating.

One important aspect of developing stronger public input and participation in biosolids management is training. For technical jobs such as wastewater treatment and biosolids management, the focus of training is on critical operations and maintenance of infrastructure, proper operating practices, and worker safety. Even larger biosolids management programs that have the benefit of a public affairs and/or public relations person may not have staff trained in the concepts and strategies recommended for effective, constructive public input and participation. Recognizing this, a workshop was held April 11, 2002, as part of this research project, to provide training to project advisors, including biosolids managers and other stakeholders from around North America. Dr. Lawrence Susskind, Professor at the Massachusetts Institute of Technology and lead author and editor of *The Consensus Building Handbook,* provided basic training on the "mutual gains approach" to addressing public conflicts and "How to Work with Your Critics." The mutually agreed to summary of this workshop is included as Appendix D on the accompanying CD-ROM.

The following is a compilation of recommendations developed from the annotated bibliography, literature review, case studies, survey, and workshop. It provides conceptual models for understanding the development of current public perceptions, including the influence of many factors—technical issues, typical human responses, communications, information, politics, and social context.

The information presented is intended to help biosolids managers better understand the social aspects of biosolids management. It provides insight into how and to what the public reacts and why, how conflicts evolve, how to manage conflicts if and when they occur, and how to minimize (and ultimately avoid) program-disrupting conflict in the future. In general, conflicts appear to develop because of differences in people's perceptions, goals, and values. People have

---

The term "public" applies to all citizens—a broader group of people who may or may not be directly involved or impacted.

[2] The term "best management practices," as used in this report, refers to a variety of actions that those managing and using biosolids take to help ensure that any potential risks or negative impacts from their biosolids recycling program are reduced as much as possible. Best management practices ("BMPs") develop over time, through general agreement within the field, and are usually based on research findings. Occasionally, researchers, regulators, or agricultural advisors (e.g. cooperative extension) compile into a guidance document current best management practices for a particular aspect of biosolids recycling (e.g. EPA's *Guide to Field Storage of Biosolids*). BMPs develop and improve over time. High-quality biosolids management is knowledgeable about—and utilizes—the most current BMPs available.

differing needs and wants that inevitably create conflict—and in a democratic society, people expect to speak up and defend their interests. Biosolids managers must be willing to join the fray, respectfully championing the interests of their customers, their utilities, their biosolids management programs.

1.1.1 Cautionary Notes

The conceptual models and recommendations presented in this chapter are based on analysis of past biosolids management experiences and social science research regarding similar environmental public policy issues. These models and recommendations have been informally tested by the authors. Similar recommendations have been put to use in other environmental fields, with considerable documented success. Nonetheless, the recommendations below may benefit from additional formal field testing and evaluation for effectiveness within the biosolids management field.

The models and recommendations below are not guaranteed to lead to any particular outcome(s). However, the authors believe—and experience indicates—that utilizing these concepts and recommendations will increase the likelihood of attaining more stable, sustainable, cost-efficient, and environmentally beneficial biosolids management programs that enjoy public support.

At the same time, biosolids managers should recognize that "the public" in their area may include a few dedicated critics of biosolids recycling who will take advantage of any opportunity to further their particular goals. It is important to involve these outspoken critics, listen carefully to their concerns, and take actions to address those that make sense. At the same time, however, it is essential to work with a diversity of other public representatives in order to obtain a broader view of public perspectives and craft a biosolids management program that a large proportion of the community can support or accept.

1.1.2 A Focus on Land Application

The following discussions often refer to "biosolids land application." This term is commonly used to identify biosolids recycling programs that involve application of bulk, Class B biosolids (or bulk Class A products) to farm soils or lands being revegetated after disturbance. Other forms of biosolids recycling involve more highly processed and stable Class A general-use products, such as composts and pellet fertilizers. Land application of Class B biosolids has been, by far, the greatest focus of public criticism. One reason is that, historically and presently, the majority of recycled biosolids in the U.S. and Canada are Class B (or equivalent). The other reasons are that Class B biosolids are usually less stable, can be more odorous, and contain viable pathogens.

The focus of this chapter, therefore, is on land application issues that develop around farms and reclamation sites where bulk Class B (or some bulk Class A) biosolids are used. Nonetheless, many of the recommendations apply as well to Class A biosolids recycling, and the overall approach recommended should prove useful in all biosolids recycling efforts.

What may be required, however, of the biosolids manager or public agency, is an evaluation of the costs of implementing robust public outreach and participation in a particular program. Many Class A compost or pellet fertilizer programs will have different public outreach and participation needs than a Class B land application program, and the costs will be different. The project team estimates that most programs that currently recycle Class B biosolids can benefit greatly from strengthening public outreach and participation, and that such programs will find the costs, while sometimes considerable, to be worth it.

1.2 What is the Problem?

Many managers and staff of biosolids recycling programs can relate to "A Bad Day in Biosolidsland."

1.2.1 A Bad Day in Biosolidsland

It has been almost a year since you started the paperwork, mapping, testing, site evaluation, a public hearing, and obtaining the permit. Finally, the trucks are actually rolling, carrying biosolids to the field! The tractors are spreading it. Hurrah! You're happy, and the farmer is happy. Now the biosolids are finally being applied so the corn can be planted. Your hard work has paid off.

But then, late in the day, you get a call from a local politician asking questions about the biosolids. They've been getting calls from worried neighbors complaining about unusual odors. You remind the politician that you had met with them about this some months ago. Yes, he says, but we didn't know it was going to upset people. Within a few days, many neighbors are angry at you and angry at the farmer. A local opposition group is formed. Your biosolids program is suddenly front page news in the local paper—and not good news. Within weeks, a public hearing is being held on adopting a local ordinance to ban the local use of biosolids. Even the politicians are voicing opposition. It's happened so fast, you don't even have time to defend your good work. What happened?

The conflicts that develop around biosolids recycling are uncomfortable for everyone. People become emotional. People get angry. Biosolids managers feel defensive, hurt, indignant, right, wrong, uncertain. For many people, this is their first experience with public conflict, and it doesn't feel good.

The difficulty for the biosolids manager is that there are many conflicting perspectives about biosolids recycling. And there are many stakeholders—the farmers using the biosolids, their neighbors, researchers and agricultural advisors, local politicians, local and regional environmental groups, state regulators, federal regulators, and more. All have their own understanding, perceptions, and interests in biosolids recycling:

- the agricultural advisor is concerned about farm economics, crop health, and overall regional soil and water quality;
- regulators feel pressure from all sides (and some say they have done a good job if everyone is equally mad at them);
- the farmers are concerned about biosolids quality and need the biosolids at specific times;
- the environmental groups are concerned about long-term impacts of trace biosolids contaminants on soils and ecosystems.

While much of the wastewater treatment process requires little interaction with the general public, biosolids recycling is different—it extends wastewater operators' work into communities and the consumer marketplace. Biosolids recycling program managers need to look at the "social" factors that affect a biosolids recycling program. If they don't, they will likely end up, sooner or later, getting some surprising and unpleasant feedback from the public.

1.2.2 Why Bother?

As biosolids managers face their first round of public opposition, they almost always ask themselves, at some point, "Why bother? Is it worth this hassle?" The stakeholders in a biosolids recycling program are on a continuum with regard to their level of commitment to the program. So when the question "Why bother?" comes up, some will throw in the towel more quickly than others.

Private contractors or consultants, whose business is developed around the concept of biosolids recycling, know that they must continue to work through any issues in order to stay in business. They will often work the hardest to protect the biosolids recycling program.

Public utility staff and management are often less motivated because their livelihood does not depend on recycling biosolids. They can work within their agency to develop other options for their biosolids, such as landfilling or incineration, and they get paid just the same. (Generally, those other options meet little public resistance, because they keep biosolids out of sight and out of mind in controlled disposal systems.) However, wastewater utilities may have a conviction that biosolids recycling is the most environmentally sound form of sewage sludge management, and that conviction can motivate commitment to the program. In some areas, utility managers know that biosolids recycling is a lower cost option, which benefits their ratepayers.

Farmers or other users of biosolids also vary in their commitment to the biosolids program. There are those for whom the cost and time savings are considerable and they are motivated and willing to defend the program vigorously. But, other farmers and users of biosolids do not realize—or may not recognize—such significant benefits. They are less willing to extend their already long work hours, less willing to attend emotional meetings where they stand accused, and less willing to stake their reputations in their local community.

In many cases, the result is that a biosolids recycling program gives up and goes away before the public conflict reaches any resolution. The decision to give up is often arbitrary, based on the level of discomfort felt by the particular biosolids program manager, staff, or farmer.

In the end, the defense of biosolids recycling often falls disproportionately on the private contractor or consultant. Because of this, a public perception is generated: the private contractor ("who is interested only in profits," cynics will say) and the public utility biosolids manager ("interested in curbing costs") are the ones who stand before the public and most vigorously defend biosolids recycling. State regulators, who are subject to public pressure, don't want to appear to promote a practice that they also must regulate (Beecher, 1999b). And because they are often not involved in discussions (because of the long hours they work), farmers or other users of biosolids products are presumed, by cynics, to be "duped" by the private interests. So a strong public perception has developed that biosolids recycling is mainly the interest of the private companies who make a profit by it. The 2002 survey for this research, discussed in Chapter 4.0 (see accompanying CD-ROM), and experiences around the country, have shown that if the skeptical public perceives a profit motive at work, their level of skepticism and concern increases.

Ideally, the question "Why bother to recycle biosolids?" should be a larger policy question that is addressed by a community or region or society as a whole, including the wide variety of stakeholders (including concerned citizens and environmental advocacy groups).

Absent such a process, and if public concerns continue (which seems likely into the foreseeable future), more biosolids managers will ask "Why bother?" and give up on biosolids recycling.

1.3 What's a Biosolids Manager to Do?

1.3.1 Building Public Participation Will Help

The answer to the sometimes negative public perception (and sometimes public anger) facing biosolids management programs today is to build public participation and earn public trust.

1.3.1.1 Don't Make Decisions in a Vacuum

Fortunately, biosolids management is not something that anyone is required to do on his or her own. Biosolids managers do not have to make decisions in a vacuum. In fact, if they want to ensure that the decisions being made are the best possible—and the most supportable—they should ensure that they have input from a variety of stakeholders.

Ideally, such input would start at the beginning of the process, when a utility and municipality are deciding what to do with the sewage sludge generated locally. There would be a multi-stakeholder group, including representatives from local and regional environmental groups, community leaders, and people from neighborhoods that are near a possible land application site or facility. Such a group would have the best chance of devising a workable plan that would be defended by a wide variety of stakeholders.

Of course, it is rare that a biosolids program is at that point—many are operating already. A biosolids program that is operating can, however, start involving the public more at any time. Convene a group of diverse stakeholders. Have them address the questions: "How is our program doing? How might we make it better? How do you envision our local biosolids being managed in five to 10 years? What might we be working towards?"

By involving a variety of stakeholders, the biosolids manager and the wastewater treatment utility as a whole share the burden of responsibility with the community.

1.3.1.2 Come Out From Under the Radar

Biosolids management is a public service, provided—directly or indirectly—by public agencies. Today there is a society-wide expectation that public programs should have significant public input and participation (Renn et al., 1995). Even if citizens do not express any interest in participating or providing input—as is often the case with wastewater and biosolids management—they still expect and feel it is their right to give input and participate whenever they choose.

In stark contrast to this societal expectation is the common practice amongst wastewater and biosolids management and staff to operate "below the radar." "Keep the biosolids program hidden," they say. "If no one is noticing and all is quiet, that's a good thing." This attitude is driven by:

- fear of public anger if the program is noticed;
- fear of media attention that often stresses the negatives; and
- farmers' parallel desire for staying "under the radar," because of concern that their crop values may be reduced by negative public perceptions of biosolids.

These are all legitimate fears. But hiding and hoping to avoid being noticed is not the answer. Yes, that approach has worked for some biosolids recycling programs for some time. But it is risky: once "discovered," such programs can face debilitating public outrage—like the "bad day in Biosolidsland" scenario above.

Today, a small but growing percentage of the public is learning about and becoming interested in biosolids recycling, most often because of complaints about odors and the growing perception of potential health effects. When public interest is initiated by complaints, there is a negative start to the public interest and involvement.

It is wiser for the biosolids manager to show off the program, to proactively seek public interest and involvement. Get people to notice the program when it is going well, so that there are some positive public perceptions of it to counteract the negative impressions that might be generated by the inevitable occasional problem. This chapter discusses how to do this.

1.3.1.3 "But There are Critics Just Waiting to Attack!"

Today, any public discussion of biosolids recycling at the local, regional, or national level may draw the attention and involvement of vocal critics. The biosolids manager's best efforts at "telling the biosolids story" may be threatened by public concerns fueled by:

- communications that are faster than ever (e.g., via the Internet);
- organized national opposition;
- occasional media stories reporting on conflicts over biosolids (when people read reports that include conflict about something, their concern about that thing increases—Covello and Sandman, 2001);
- heightened public sensitivity to environmental issues;
- the public's tendency to oppose anyone *perceived* as threatening the environment or public health ("the polluters"), especially "industry," "corporations," and "bureaucratic government;"
- increased public fears and anxiety about all kinds of perceived threats (Freudenberg, 1996).

On top of all these, there are more biosolids to manage and fewer remote areas in which to manage them—so biosolids programs are increasingly close to the public.

Biosolids managers have to accept these circumstances. This is the playing field on which their biosolids programs must play, and broad societal expectations are generally out of their control. What biosolids managers can control is how their biosolids management program acts and reacts within this social context.

1.3.1.4 There is Hope ...

Despite the apparently momentous hurdles facing biosolids managers, there is much that is reassuring. For example, social science researchers have noted that in many public debates, there typically is a small percentage of the public (maybe 10%) that will vocally oppose a proposed project and another small percentage that will vocally support it (again, maybe 10%).

The remaining majority of the public (maybe 80%) does not participate much but, if the debate becomes highly visible, will listen to both sides of the discussion. These quiet, undecided people have been called "guardians" by some (Ozawa and Susskind, 1984).

"Guardians" will tend to believe and side with the opposition if they perceive the proponents of the proposed project as being arrogant, condescending, close-minded, unsympathetic, dishonest, distrustful, and/or unfair. Other research has shown that many people make decisions about whether something is good and safe based not on technical information as much as on the perceived trustworthiness of those providing information about it.

So biosolids managers have an opportunity to avoid turning the guardians—that critically important majority—against them. To take advantage of this opportunity, biosolids managers must consider carefully how their words and actions are perceived. Does it help to operate "under the radar" or is that perceived as having something to hide? Does it help to tell someone you are the expert and they shouldn't worry—or is that perceived as condescending?

Wouldn't you like it better if someone explained to you in advance that some unfamiliar trucks will be dumping material on the neighboring farm field and that if you have any questions he or she is available to talk with you? Wouldn't you like it better if someone listened carefully to your concerns and expressed interest in working to address them? If biosolids managers listen, develop dialogue, and work with neighbors and other stakeholders, they will earn the trust and support of many of those stakeholders—as well as the "guardians."

As described in the case studies in Chapter 3.0 (see accompanying CD-ROM), there are reassuring examples of biosolids recycling programs that have benefited greatly from public input and participation—gaining increased public trust and support, reducing risk and vulnerability, and improving the long-term sustainability of the program.

1.3.1.5 Taking the First Steps...

As a biosolids manager takes the scary first step of bringing their biosolids program "out from under the radar," they initially find that no one seems interested. They announce tours, they hold meetings—and no one shows up. Why? Everyone is busy, and biosolids are not that interesting to most people. Then the vocal critics appear—and suddenly public interest begins to grow, because there is controversy.

Early in this process, hopefully before any controversy starts, is the time to demand help. Biosolids managers carry out a significant responsibility for the public and they should be able to expect the public—at least representatives of public interests (e.g., politicians, community groups, universities, cooperative extension, and non-profit environmental advocacy groups)—to help them with their public responsibility for biosolids management. While ultimately it is the biosolids managers' responsibility to manage biosolids using their best professional judgment, the public and political and environmental leaders are expected to be responsible and not just criticize. The public should help guide biosolids management decisions through constructive, collaborative problem solving. Of course, the biosolids manager also has to provide opportunities for public involvement that work for the public, such as holding evening meetings in neighborhoods rather than daytime meetings at the plant.

The ideal goal of public participation is to get to the point where the biosolids management program, no matter what type of use or disposal it involves, has the support of informed and involved individuals in the communities in which it operates. Ideally, these people have easy access to information about the program and are willing and able to pick up the phone and ask questions or make observations. If they note something that could be improved, they are comfortable making suggestions and the biosolids manager welcomes their input. The biosolids manager and staff check in with these people now and again to be sure the relationships are healthy and communications are good. They are all part of a team, working together to make the biosolids management program as beneficial as it can be.

In summary, public scrutiny is a part of our open society, and there is always a risk of "the public veto." Why ignore that risk? It can be argued that it is irresponsible for an agency manager to put millions of public dollars into any engineered solution without a process of public participation in the development of the project. There is just too great a risk of wasting money should the project fail from lack of public support.

1.3.2 "Public Acceptance" Versus "Public Relationships"

The literature on the interactions between biosolids recycling programs and the public has long been focused on the concept of "public acceptance." As the team working on this project began collecting information in 2000, this was also the focus of this project. For example, the first draft of the literature review (Chapter 2.0—see accompanying CD-ROM) was imbued with the viewpoint, held by many biosolids managers, that interactions with the public are aimed at gaining acceptance of existing biosolids management practices. The biosolids manager, in consultation with the biosolids user, regulatory staff, and agricultural advisors, makes the decisions. These are the technical experts who know the science. Once the decisions are made, the public just needs to be educated, so they will accept the decisions. This "decide—announce—defend" ("DAD") approach historically has been the most widely used approach in biosolids management. Not surprisingly, the management of wastewater treatment byproducts is not high on the list of things that most members of the public think or care about. So it has been easy for biosolids managers to operate without much public interest or participation.

During the course of this project, however, the project team underwent a change in perspective. By looking at the bigger picture—observing the trends in biosolids recycling public acceptance and recognizing that biosolids recycling conflicts are similar to other environmental conflicts—the shortcomings of the "let's gain public acceptance" approach became apparent.

1.3.2.1 Characteristics and Outcomes of the "Public Acceptance" Approach

Efforts to gain public acceptance have been discussed in numerous biosolids management industry conference papers on "educating the public." The focus has been on conveying the information the experts have, in order to convince the public. As has long been hypothesized (Logan, 1995; see also the survey results in Chapter 4.0), when citizens have only a little information about biosolids recycling and its potential and perceived risks, they may experience a large level of concern, even fear. Efforts to develop public acceptance through conveying information often go only as far as getting into the average citizen's mind this minimal level of information. Then the public reacts from their concern and fear. As noted below, information and communications are critical components, but not if they are only aimed at convincing the public you are right.

The "let's gain public acceptance" approach encourages setting goals and expectations that may be dangerously out of sync with known public perceptions of biosolids recycling. The survey conducted for this project (see Chapter 4.0) detected a significant, apparently inherent uneasiness with the concept of biosolids recycling. Given this fact, providing only minimal information and outreach to the public about biosolids may backfire.

Think of the typical opportunities that many biosolids managers use to provide information to the public:

- a single public meeting or hearing,
- a brochure and a fact sheet,
- a letter to abutting landowners,
- a notice in a local newspaper,
- an information-packed website.

All but the first of these are one-way communication tools. Even a public hearing too often becomes a lecture by the experts with just a few questions from the audience. The biosolids manager has no opportunity to hear the diversity of thoughts and perspectives in the community. He or she only hears a few vocal opponents, and they do not seem credible at all. So the biosolids manager can write off public concern, because the concerns heard are ridiculous. But the public is diverse! A few vocal critics do not speak for the numerous other perspectives at a hearing.

If biosolids proponents use only the traditional tools listed above, then, when questions start coming at a public hearing, there are no arrangements for further dialogue, for understanding the people's perceptions and concerns. Instead, the biosolids manager feels cornered and defensive. He or she tries to give more information to alleviate the concerns. But, at times like this, people do not like to be talked at, talked down to, and/or pressured to be convinced of something. Piling on more technical information at such times is a mistake. People don't like to have their opinions, concerns, and questions rebuffed, put down, or ignored. They like to take in the variety of perspectives and information and mull it over before they make any decision. But often the biosolids manager's goal is to get through the public hearing as soon as possible so that the needed permit is granted ASAP. Pushing forward with an agenda at a time like this can incite public anger. As they try to push their way through the hearing, proponents of the biosolids program lose credibility, trust, and respect. It's at this point that you begin to hear angry personal attacks, like "You're not from here. You don't care about this community. You're just out to save a quick buck by dumping your filth on some poor rural town!"

Even worse, the depth of this anger occasionally motivates some individuals to devote days and days of their lives to unpaid work opposing biosolids recycling. The devotion to that work, in some cases, may be driven, as much as anything, by a motivation to inflict retribution on particular individuals or on the "biosolids industry" as a whole, which they view as arrogant, condescending, uncaring, and money-grubbing.

In the "typical bad day in biosolidsland" scenario outlined above, think through what the neighbor's reaction might be to the "public acceptance" approach and what that could lead to (we'll call the neighbor Sue and the biosolids manager Jack):

> *After a weekend of putting up with some periods of bad odors, Sue finally reaches the biosolids manager, Jack, on Monday morning. They talk.*
>
> *The biosolids manager speaks and acts on the assumption that Sue just needs to learn more about biosolids and then she will understand the minimal risk and calm down and accept the situation (the information-sharing, "public acceptance" approach). So Jack starts explaining the science behind the practice. He is surprised when Sue interrupts him with*

some anger in her voice: "What's in that stuff? Is it harmful?" His efforts to provide factual technical information and calm clear answers do not have the intended effect. But he does not recognize this. Instead, he continues with more information. At the same time, he begins to feel tense and defensive. "Why is she attacking me? I do excellent work and this is good for the environment."

Jack has not recognized that, in many cases, if someone takes the time and effort to pick up the phone and call, there is a fairly strong motivation for doing so. People don't bother to call their wastewater treatment facility for the fun of it. There is likely to be emotion behind such a call. Jack failed to find out what Sue is thinking and feeling and address that.

Sue's frustration comes from her natural uneasiness with something different nearby that doesn't smell good. She feels a loss of control—something is happening to her and she can't do anything about it. Add to that the fact that she hasn't been able to find out any specific information for more than two days. And now this biosolids manager is talking to her in technical jargon that does not seem to address her experience and concerns. People want their concerns heard and validated. Jack's well-intentioned effort—"you just don't understand the science; let me explain"—is off target. Sue sees his effort as unsympathetic and patronizing. Her frustration builds. Her fears increase. Jack is losing credibility. She begins not to trust him.

Further, if Sue happened to have gotten information from the Internet, or from established opponents to biosolids recycling (and Jack can't know whether this has happened or not—unless he asks Sue questions and listens carefully), Sue's distrust grows: Jack is clearly not telling the whole story.

Remember that typically the neighbor knows almost nothing about biosolids, and does not have the same context the biosolids manager has: soil science, agriculture, wastewater treatment, etc. So the information Jack provides is meaningless: at this stage, the neighbor can't judge whether something makes sense and whether it is "true" or not.

Further, if Sue has heard back from a dedicated opponent who expressed sympathy, the unsympathetic Jack has no chance. As Peter Sandman (2000) points out, active opponents—e.g., the dedicated opponent who returned Sue's call—do not create public outrage, they merely "harvest" it. The concerned neighbor wants to believe someone, to be reassured that her experience was real.

So Sue makes a judgment based on her perceptions of the circumstances and the people she has communicated with:

> *"This stranger, Jack, who has a profit motive, is telling me this stuff that disturbed my weekend is okay. But my experience is that it stinks. And I have read on the Internet that it is not okay. I talked to another stranger who was empathetic to my concerns and validated my experience. I don't really know enough about this biosolids stuff*

to know whom to believe. Meanwhile, this biosolids person goes on talking to me in techno-babble."

Who would you be more inclined to trust, if you were in Sue's shoes? Someone who you perceive as responsible for creating the uncomfortable experience and who is unresponsive to you? Or someone who empathized with you and validated what you experienced?

When a problem arises—an odor issue, a neighbor's concern (and it is almost inevitable that problems happen now and again)—public ignorance or grudging public acceptance of your biosolids program will not help you. What is needed at such times is a high enough level of local public trust and established public relationships so that communications and exchange of information remain possible, even when emotions run high. Wouldn't it be helpful if the biosolids manager, Jack, could say to the concerned neighbor, Sue:

"I understand that you are feeling concerned about the biosolids on the neighboring farm. You might want to talk with others. For instance, your neighbor, Joe, was involved in reviewing our biosolids program. He had some of the same concerns. And we have a meeting scheduled at your neighborhood meeting place on Tuesday evening, because we expected there might be some questions, once the biosolids were applied."

Building public relationships can provide many rewards over the long term.

1.3.2.1 Characteristics of the "Public <u>Relationships</u>" Approach

The new perspective gained through this project attempts to find ways to avoid the pitfalls of the historical "public acceptance" focus. This new perspective, developed from understanding public perceptions, conflict, and conflict resolution as experienced in numerous other fields, aims to enhance the integrity and credibility of biosolids managers and their public relationships with diverse stakeholders.

As discussed extensively below, biosolids managers are encouraged to cultivate constructive relationships through enhancing two-way communications skills. Listening, being open to others' ideas and perspectives, and learning from constructive criticism can enhance the biosolids manager's public relationships. This process also provides important understanding of the public perceptions and social context within which a biosolids recycling program operates. Gaining public input and participation also provides critical information to decisions, so those decisions will end up being better and the program more sustainable.

The practices that people learn to reduce tensions, heal conflicts, and build long-term relationships in their private lives also can be applied to public relationships. Think of how you feel when you are in a situation like the neighbor described above. Think of what you would want the other person to say and do:

"I hear your concern. I apologize that we caused you some discomfort. We are addressing the problem in this way. We appreciate your patience and want to avoid you having the same discomfort again. We'd like to talk more with you—get your input. May I call you tomorrow to update you on our progress? Here is my phone number, in case you have any questions. Call anytime."

The authors hypothesize that excellent, long-term biosolids programs have been successful because of the public relationships around them (such success has often come to these programs without people realizing that the building of public relationships was what made the difference). The public relationships have been more valuable than the fact sheet, m the technical report, or the letter to the editor. Hosting a tour is the best example of a simple relationship-building tool; it makes a difference because the public becomes aware that the utility has nothing to hide and that the people who work there are fine people. Open information on a website makes a difference because it shows the utility is transparent and open. The odor complaint hotline makes a difference because it shows that the utility is sensitive to an issue that affects the neighbors and welcomes anyone's input. The common denominator in the success of these efforts is that they build better relationships with a variety of people.

Based on the results of the literature survey, case studies, and public survey, the authors recommend a "public relationship" approach to biosolids recycling in place of the "public acceptance" approach. This approach agrees with what social scientists, who study conflict around environmental issues, recommend. It's not "public relations;" it's "public relationships." Critical components of this approach are briefly outlined here. Many of the same concepts are discussed in more detail later in this chapter:[3]

- TRUST AND FAIRNESS: Constructive public relationships are built on trust and fairness. Biosolids managers build trust by repeatedly doing what they say they are going to do and communicating a lot with people who may be affected. Fairness is worked out through direct and honest negotiations with stakeholders. Neither requires that people be best friends, or even like each other. Public relationships are not necessarily about friendship and enjoying each other's company. They are about getting along decently, respectfully, and equitably in the particular situation—even when there is strong disagreement.

- INFORMATION AND COMMUNICATIONS: Public relationships are facilitated by good information and communications. Biosolids managers can ensure that the information they provide is clear, direct, complete, and useful to the recipient and that the biosolids program and staff provide clear verbal and written communications at all times. Remember, it's important to be communicating during non-conflict times, as well as during times when there is tension. Good information and communications do not necessarily take lots of time and energy; just remembering to make a call or send a note before the trucks roll down the street can help. It's a matter of courtesy, showing that you understand and respect others' needs and concerns.

- ORGANIZATIONAL COMMITMENT AND MOTIVATION: Public relationships, like good biosolids management programs, are built over time and require strong organizational commitment and motivation. Throughout the literature on biosolids public

[3] The components presented here are consistent with the developing Water Environment Research Foundation (WERF) conceptual framework of "Public Partnering." These concepts—trust, fairness, information, communications, etc.—were identified from the literature during the first year of this project. They were also identified independently by a parallel WERF project regarding public perception and participation in water reuse (Hartley, 2003). They were further refined through discussions with that project team, including especially Troy Hartley, Ph.D., Principal Investigator, Juliana Birkhoff, Ph.D., and Mary McDaniel, M.D., J.D., MPH. The authors extend our appreciation to each of them.

perception and acceptance, there is constant encouragement for agencies and companies to commit adequate resources to working with the public (CH2M Hill, 1982; Hartley, 2003). Unfortunately, all too often, the public participation or public education or communications line item is the first to go when budgets are trimmed. The authors believe that biosolids recycling may not remain a long-term viable option for organizations that fail to provide significant, on-going commitments of finances and personnel to developing public outreach, information, communications, and public participation. If such a commitment is not made, it may be best to manage biosolids in some way other than land application, so as to minimize interaction with the public.

- OVERSIGHT AND ENFORCEMENT: Today, across North America, one of the largest issues around biosolids recycling programs is the adequacy of oversight and enforcement (U.S. EPA Office of Inspector General, 2000; National Research Council, 2002). Public knowledge, understanding, and trust of biosolids management programs are enhanced by direct public observation and monitoring. Biosolids managers can provide opportunities for stakeholders to see how biosolids are produced, treated, managed, and utilized in compliance with local (if applicable), state, and federal regulations. Because biosolids recycling is highly regulated and *its safety depends on strict management and compliance*, the public's confidence will be increased if they know that biosolids are being recycled properly. Thus, biosolids programs will benefit by ensuring independent monitoring and by openly and proactively providing detailed records of biosolids recycling activities. Biosolids managers are encouraged to involve stakeholders in designing and implementing local oversight and monitoring systems or witnessing existing oversight and enforcement conducted by state or local authorities[4].

While the authors recommend that biosolids managers develop strong public outreach and communications programs, they see such programs as only a part of a larger public participation effort. Public outreach, or public relations, without public participation (e.g., two-way communications and public input in decision making) is not enough and may result in public outrage.

Ask yourself these questions as you decide what to do or say in any given situation:

- Have I listened enough and am I sure that I have all critical information?
- Do I understand the situation from a variety of perspectives, including community and public stakeholder perspectives?
- Have I covered all the angles and am I sure that I have incorporated diverse perspectives and interests in this decision, this action, or what I am going to say?

Be humble—you don't have all the answers. Then, ask one last question before you take any action or make any statement:

"What will be the impact of my next action on our program's relationship(s) with key stakeholders?"

If the impact(s) will be negative, consider whether or not to continue with that action.

---

[4] These concepts fit well with the National Biosolids Partnership's Environmental Management System for Biosolids program (National Biosolids Partnership, 2000b)).

How critical are the relationships that will be negatively impacted? Remember that if trust is broken, it can take years to regain. Are there other actions you can take, such as talking to the impacted stakeholders in advance, that will help lessen the negative impact on the relationships? You will, at times, have to take actions that will be unpopular with some people—but it is best to do so knowing what the reaction will likely be. And, if possible, it is always best to let people know in advance what you are doing and why.

1.3.3 Benefits of Public Participation Done Right

As noted above, biosolids managers benefit from public participation by knowing that someone else supports their decisions and actions; managers don't have to make decisions alone, in a vacuum. The most significant benefit of public participation in biosolids management decision-making and implementation can be that it leads to better decisions and improved environmental performance and efficiencies. By working with a variety of public stakeholders through a robust public participation process, biosolids managers may ensure that decisions are made with as much information, and with as many different and specialized perspectives, as possible. With an able group of well-informed stakeholders—including agricultural advisors, communications specialists, local environmental group representatives, farmers, local politicians, neighbors, engineers, and wastewater operators—the biosolids manager will:

- decrease the likelihood of making mistakes,
- reduce the vulnerability of the program to criticism, and
- find a more efficient and environmentally beneficial solution.

However, it must be noted that there is no guarantee that public involvement will result in the choice of the best possible solution—that depends on how careful, fair, and equitable the public input process is. (There are many examples in the literature regarding public participation in environmental decision-making that resulted in poor and wasteful choices.) And it is certainly possible that a robust public participation process may result in a decision that is not the initial first choice of the biosolids manager. If it's a good choice, that's okay. Ideally, the final decision can be one that everyone is able to accept, even if it may be no one's first choice (this is what is often achieved by consensus). To avoid bad choices, however, the biosolids manager—and all other participants in the decision-making process—must be willing and able to strongly state his or her interests. Experts need to state clearly their best professional judgment and to deny any decision that they know is not technically viable, environmentally sound, or economically possible. If needed for critical and potentially costly decisions, biosolids managers can learn about and/or seek assistance to create a robust public participation process that will avoid potential pitfalls.

1.4 Stumbling Blocks

Biosolids managers want to know "What works?" That is what the authors attempt to answer in the rest of this chapter. But before turning to that, it's good to review a few things that *don't* work.

Some actions and attitudes of biosolids managers will increase the likelihood that a biosolids recycling program will not be sustainable for the long-term. Over the course of this project, the following "stumbling blocks," or factors, were found to impede public participation and the development of public trust. Although some of these factors cannot be controlled by the biosolids manager, the way in which the program addresses these issues is something the biosolids manager or public agency can control.

Stumbling blocks that are largely out of a biosolids manager's control include, but are not limited to:

- public uncertainty about the concept of recycling materials that are derived from sewage or human waste;
- widespread public environmental worry about possible hazards of wastes, especially those that include industrial sources;
- public uneasiness about anything "new" (even though formal biosolids recycling programs have been going on for decades, they are "new" to each individual as they first encounter the concept);
- rural citizen distrust of urban products, especially urban "wastes";
- a low level of general public knowledge and understanding of wastewater and biosolids management;
- the public perception (and, in some cases, actual reality) of little or no enforcement or oversight on the part of local, state, or federal government officials (or other form of independent monitoring and oversight).

Stumbling blocks that are in biosolids manager's control include, but are not limited to:

- a poorly designed and operated biosolids management program (proper siting is critical!);
- use of biosolids that create highly negative personal experiences for neighbors or other members of the public (e.g., poorly controlled odors);
- biosolids managers, staff, and biosolids end-users who are *not* knowledgeable, respected, open, and honest;
- poor communication (especially lack of listening);
- ineffective communicator(s);
- lack of credible information, or conflicting information;
- pre-existing conflicts between biosolids end users / program participants and neighbors / communities; and
- attempts to cut corners to save money (a situation that may be encouraged by the competitive marketplace for sewage sludge management services both private and public).

The following sections provide further discussion of what works and what does not work in addressing the social aspects of biosolids recycling. Many of these recommendations are not new (see, for example, CH2M Hill, 1982), but their consistent implementation by biosolids managers has not yet been realized.

1.5 Twelve Steps for Developing Public Participation and Earning Trust

Based on an evaluation of how a biosolids recycling program is currently perceived by a variety of stakeholders, managers can develop goals and strategies to improve long-term relationships with various stakeholders. Recommended actions for developing public participation and earning public trust are distilled into twelve critical steps (Table 1-1). Each of the steps is discussed in more detail in the following sections.

Small biosolids management programs may find a less detailed approach more suitable, but the overall concepts will apply. For example, even small programs will benefit from building organizational commitment to public participation (Step 1), identifying and getting input from stakeholders (Steps 5 and 6), providing a range of public participation opportunities with a focus on opinion leaders (Steps 8 and 9), and making public participation an ongoing effort (Step 12). These most critical action steps appear in bold in Table 1-1.

Table 1-1. Twelve Steps for Developing Public Participation and Earning Trust.

Step 1.	**Build commitment within your organization.**
Step 2.	Understand the national and local context.
Step 3.	Learn more about conflict.
Step 4.	Determine what is beyond your control; work on what you can control.
Step 5.	**Determine who your stakeholders are.**
Step 6.	**Get input from key stakeholders as soon as possible, early in the process.**
Step 7.	Redefine success.
Step 8.	**Build public relationships.**
Step 9.	**Improve communications, especially listening and dialogue.**
Step 10.	Provide information useful to *recipients*.
Step 11.	Don't be afraid of conflict—work patiently through it.
Step 12.	**Continue to monitor public perceptions and public relationships.**

1.5.1 Build Commitment Within Your Organization

*** a *most critical* action step**

How committed is your organization to public outreach and participation? Successful public outreach and participation requires funding and attention. This means that biosolids management budgets and contracts need to explicitly include staff time and other costs associated with public outreach and participation. Absent this dedicated funding and staff time, biosolids programs may be wise to select management options that impose as little as possible on the public. For example, consider only very isolated sites for Class B land application, invest in producing a fine quality Class A biosolids product that will win consumer confidence in the marketplace, or choose landfilling or incineration. Any biosolids recycling program that fails to commit adequate resources toward public outreach and participation is doing the entire biosolids management industry a disservice and may bring public outrage upon itself.

The viability of most biosolids recycling programs hinges on the ultimate cost, often measured as the cost for use or disposal of each wet ton or gallon of biosolids. Public outreach and participation costs are not generally included in budgets as part of the basic cost of biosolids management—they are seen as added costs. In many communities, public wastewater utilities are required to follow public bid processes that require acceptance of lowest cost bids. In other locations, biosolids management is done by agency staff and wastewater budgets are scrutinized by boards and politicians. Some public contracts for management or disposal of biosolids include requirements for public outreach and participation and best management practices, but many Requests for Proposals and contracts do not include such terms, or if they do, oversight does not happen to ensure they are done. Cost too often remains the sole critical factor, and when public outreach and participation requirements are not spelled out in the budget or contract and enforced, they don't happen.

The true cost of a biosolids recycling program with robust public outreach and participation can be considerable. The highly-acclaimed King County, WA biosolids recycling program has cost the County millions of dollars over the past decade, including membership support of the NBMA, university applied research, land acquisitions, demonstration sites, and staff that independently audit each land application program (see Chapter 3.2.1). Likewise, Denver's Metrogro Farm (Chapter 3.2.9) has remained in operation to a large extent because of the City's $1.6 million funding of an independent monitoring program by the U.S. Geological Survey. And Merco, the New York City biosolids land applier at Sierra Blanca, Texas, spent $3 million on research and monitoring. The cases of Boulder, Colorado (Chapter 3.2.6) and New York City's Citizen Advisory Committee (Chapter 3.2.7) are additional examples of public participation efforts that cost considerable effort and money. Even less-formal public outreach and participation efforts—including communications, information-sharing, and media relations—require a considerable commitment of time and resources on the part of the public utility and/or biosolids manager.

Is it worth it? In recent years, it appears that many biosolids managers and agencies have chosen recycling programs based on factors other than short-term costs. Considerations have included:

- biosolids recycling may prove to be more cost-effective in the long-term as landfill costs increase and biosolids products become more valuable and return income;
- recycling biosolids is "the right thing to do"—more environmentally correct than disposal;
- the natural resources value of biosolids (e.g., for soil remediation, building soil organic matter to improve moisture retention); and
- recycling biosolids is a more sustainable, long-term option for society.

Still, issues of cost often make or break a biosolids recycling program, as seen in the Sierra Blanca land application program (Chapter 3.2.8) and the closure of the Site 2 composting facility in Montgomery County, MD (Chapter 3.2.11). In some parts of the country, low landfill tipping fees make recycling of biosolids more expensive than landfill disposal, even without considering public outreach and participation costs. Even in parts of the country where landfill costs are high, such as the Northeast, the total costs of successful biosolids recycling programs, including public outreach and participation, approach the cost of landfilling.

An additional cost issue, now routinely weighed by many communities, is the indirect costs associated with policy decisions, including exposure to risk and liability. Biosolids recycling, depending on how it is done, may or may not create less exposure to liability and legal action—and this needs to be included in cost analyses. No matter what final end use or disposal is chosen for sewage sludge, a community establishes a certain amount of risk and liability, and it is hard to be certain of how risky any particular option may be perceived to be in the future. Social scientists who promote the use of mutual gains negotiation and consensus building (Susskind et al., 1999b), suggest that there are two significant cost-related benefits of involving the public in policy decisions: 1) reduction in potential risk and liability due to improved informed analysis and decision-making, and 2) avoidance of costs associated with defending unilateral policy choices, including legal and "damage control" costs.

The biosolids management case studies (Chapter 3.0) illustrate that successful biosolids recycling programs require far more skill, patience, dedication, and commitment of resources than any other form of sewage sludge management. Programs that have met with the greatest difficulties in terms of public outrage have often been those that have not ensured adequate resources for public outreach and participation and/or have cut corners (see Chapters 3.2.13 and 3.2.14). Indeed, some biosolids recycling projects that fail to gain public acceptance do so through a downward spiral driven by cost-cutting measures:

- The lowest-bid contractor is awarded the contract to manage the biosolids recycling program;
- Stiff competition creates a low bid that does not include any resources allocated to public outreach and participation initiatives;
- The low bidder cuts corners, providing little or no training to staff;
- Best management practices are curtailed;
- The need for efficiency and speediness leads to careless mistakes such as land applying odorous biosolids during a neighbor's Memorial Day picnic without having notified anyone;
- When angry neighbors begin asking questions, the staff's lack of training and knowledge regarding the scientific basis for biosolids recycling leads to charges of incompetence;
- Staff feels threatened and defensive and either reacts defensively or shuts down the program and runs away;
- In either case, constructive, informative discussion and other productive public education, participation, and relationship-building processes don't happen.

This repeated biosolids recycling scenario results in an increase in the number of citizens with a negative impression of biosolids recycling.

Most of the 16,000+ wastewater treatment facilities in the U.S. are small, with limited budgets, and the experiences of King County or Denver may not be applicable (Ross & Associates Environmental Consulting, 2000). However, there are ways to improve the efficiency of public participation programs, while providing assistance to smaller facilities. One approach is the formation of cooperatively funded outreach programs for all biosolids programs within a region. For example, the Northwest Biosolids Management Association—which involves agricultural, farmer, and academic communities, as well as wastewater and biosolids management organizations—conducts cooperative public outreach, funded disproportionately by larger agencies who understand that a problem with any biosolids recycling program, large or small, affects public perceptions of biosolids in general.

In summary, to build organizational commitment, biosolids managers have to estimate the costs and do their best to "sell" to their management and political boards the importance of public outreach and participation. It is a significant challenge. Write these costs into the routine operating budget. Write public outreach and public participation responsibilities into someone's job description and give them the time and funding to do it. Write into biosolids recycling management contracts strict requirements for implementation, reporting, and evaluation of public outreach and participation efforts. And make sure there is a robust system in place for checking that these requirements are followed.

1.5.2 Understand the National and Local Context

A review of the literature (Chapter 2.0—see accompanying CD-ROM) reveals the evolution of biosolids recycling and public acceptance in the U.S. over the past decades. Over this period, recycling rates have risen, as have efforts to research and regulate the process. Public outreach efforts and public acceptance of biosolids recycling have had ups and downs. Key developments include U.S. EPA's application of the risk assessment approach to sewage sludge, promulgation of 40 CFR Part 503 regulations, reviews by the National Research Council (in 1996 and 2002), and the October 2003 petition to U.S. EPA regarding stopping land application (see diagrams of key developments in Appendix E on accompanying CD-ROM). Because all biosolids management programs operate within the context of these national trends, it is helpful for biosolids managers to understand the historical trends and key events as they plan for the future of their programs.

One national trend has been that technical development and extensive research of biosolids recycling over the past several decades have been conducted, shared, and debated almost entirely within the biosolids field. The survey conducted for this project (see Chapter 4.0) confirmed a low level of knowledge about biosolids amongst the public. This gap in knowledge between "experts" and the lay public is common in many fields, and it has been the subject of social science research. One approach for closing this gap includes "joint fact-finding" or "collaborative research" and on-going public participation in development of understanding and policy (Susskind and Field, 1996). The Boulder Case Study (Chapter 3.2.6) successfully included this kind of joint learning process. The NBMA and King County took steps in this direction with the development of their partnership with the Mountains-to-Sound Greenway Trust (see Chapter 3.2.1). The National Biosolids Partnership has also taken steps in this direction with involvement of public environmental group representatives in the development of the Environmental Management System program.

An important result of "bringing the public along" is that, by so doing, biosolids managers will quickly come to understand when "the public" does not want to be "brought along." Thus, the biosolids manager will avoid being surprised by a public unwilling to agree to what the biosolids manager proposes.

Knowing the context is helpful in other ways in reducing the vulnerability of your biosolids program to public concerns. For example, in 2003 and 2004, there has been growing citizen concern over possible health risks from Class B land application sites. Even if conflict has not been a problem for your program, it may be worth evaluating the potential risks to neighbors of such sites and proactively managing that potential risk before concerns are raised.

However, before taking any action, biosolids managers can determine the importance of any particular issue by getting public input. Currently, many programs are adopting Class A technologies in order to reduce vulnerability from pathogen issues. At least some of these decisions have not had the benefit of stakeholder input and some programs may be surprised down the road to find that even some Class A biosolids are not accepted. With the current state of public concerns about biosolids management, any significant decision regarding a biosolids management program might do well to include public input and advice from biosolids stakeholders who are knowledgeable about current issues and trends.

A wealth of information is available to help biosolids managers understand the national and regional context in which their biosolids program operates, much of it on the Internet (Table 1-2). The National Biosolids Partnership, the Water Environment Federation, and regional associations all provide information at Internet websites and through electronic and paper newsletters. The Cornell Waste Management Institute website is an excellent source of information about current public concerns. Searching the Internet using the keyword "sludge" will identify many more sites expressing concerns about biosolids recycling. As with any information, the reader should be aware of the source of the information and its credibility.

Table 1-2. Websites with General Information and Recent Developments on Biosolids.

National Biosolids Partnership (Water Environment Federation, Association of Metropolitan Sewerage Agencies, & U.S. EPA) : http://www.biosolids.org

U.S. EPA (national biosolids program): http://www.epa.gov/owm/mtb/biosolids/index.htm

U.S. EPA (Rocky Mountain states) biosolids program: http://www.epa.gov/region08/water/wastewater/biohome/biohome.html

Cornell Waste Management Institute: http://www.cfe.cornell.edu/wmi/

California Association of Sanitation Agencies: http://www.casaweb.org

Great Lakes By-products Management Association: http://www.glbma.org/

Mid-Atlantic Biosolids Association: http://www.mabiosolids.org

New England Biosolids and Residuals Association: http://www.nebiosolids.org

Northwest Biosolids Management Association: http://www.nwbiosolids.org

State of Virginia Biosolids Program: http://www.biosolids.state.va.us/

State of Pennsylvania Biosolids Program: http://www.dep.state.pa.us/dep/biosolids/biosolids.htm

State of Washington Biosolids Program: http://www.ecy.wa.gov/programs/swfa/biosolids/

Publications that cover biosolids management news and developments; all require paid subscriptions:

BioCycle, Journal of Composting & Organics Recycling (JG Press): http://www.biocycle.net

Clean Water Report (Business Publishers, Inc.): http://www.bpinews.com

Knowing the national and regional context, it is then possible to evaluate your local program's relationships with stakeholders and the public. The first question is who knows about your program? If the answer is you, your staff, the farmer, and your family, there's work to be done. Other questions are:

- Does your program get input from stakeholders? Is that input formal—i.e., through a "complaint" hotline or an advisory group? Or is it more informal—i.e., staff talking to neighbors and the farmer every now and then?
- How much do stakeholders know about the biosolids recycling program? For example, is the farmer or end user well-informed? Could he or she answer questions of neighbors or town officials? Do program staff check in with the farmer on a regular basis? Does the farmer receive user-friendly reports on biosolids quality on a regular basis?
- How much does the general public know about it? What opportunities exist for them to learn about it?
- Has there been any conflict around the biosolids program? Has this created strong opposition that you have to contend with? Has any conflict been resolved or are there still questions or issues left hanging?
- Has the community experienced conflict over other environmental issues or siting of facilities? What was the role of the farmer/end user in that conflict, if any? What was the role of site neighbors? Often new potentially controversial proposals will re-open old wounds and divisions in communities—it's possible your biosolids program proposal could do this.
- How is the farmer's/end user's relationships with his or her neighbors? With community leaders?
- Has there been any media attention to the biosolids program? Has it been accurate, fair, and balanced coverage, or not? Have you ever spoken with local or regional media?
- What would a graph of public knowledge and support for your program look like? Is it up and down? Or steady?

These same questions can be reworded to apply to any form of biosolids recycling program or any facility site (e.g., a biosolids composting facility or pelletizing plant) or with regard to whatever communities or groups of stakeholders a biosolids recycling program operates around. Understanding your local history and context will help identify priority public outreach and participation needs.

1.5.3 Learn More About Conflict

Many of the case studies (Chapter 3.0) discuss the development of public knowledge and perception of biosolids in a particular location (e.g., see Boulder, CO and New Hampshire case studies). To develop insight into how conflicts about biosolids/sewage sludge management develop in communities, three possible models are proposed that consider the role of shared knowledge and/or shared values in the community over time.

Model 1. Conflict is triggered by a lack of shared knowledge about biosolids. When the public learns more about biosolids, their acceptance will increase.

Logan (1997) suggested that as people learn about biosolids recycling, too little knowledge can result in uncertainty and fear. If that is as far as people go in their knowledge and thinking about biosolids, then they may end up with a high level of concern and a negative impression. Logan suggests that, with more information and time to absorb it, people will come to understand biosolids recycling and accept it. Clearly, the development of strong knowledge and understanding of the technical aspects, including risk management, involved in biosolids recycling often results in an improved perception and acceptance of the practice.

Figure 1-1. One Possibility: Progression Toward Acceptance.

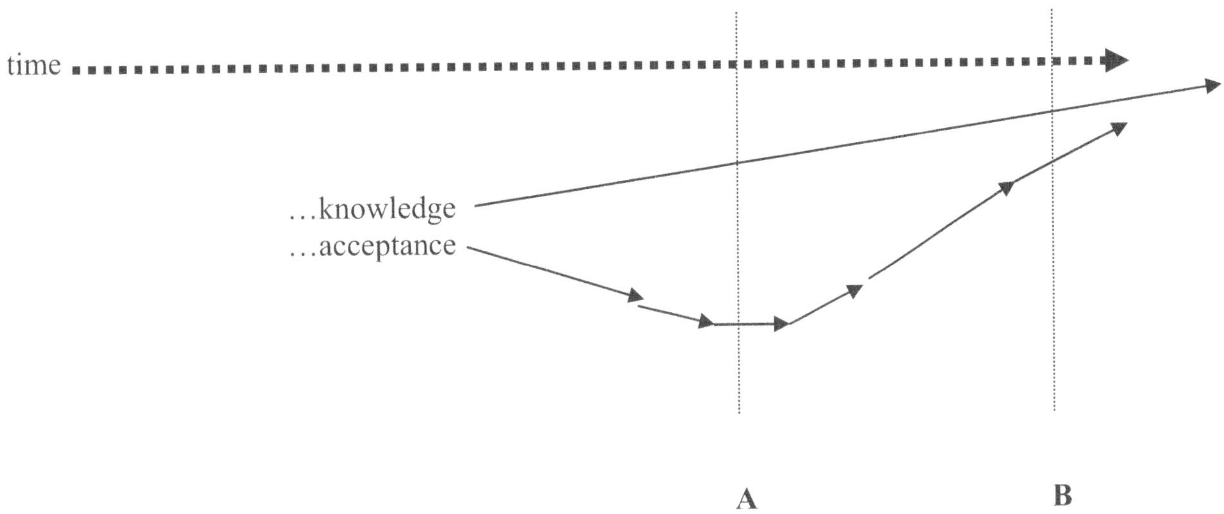

According to this model, acceptance may decline initially but will rise as the level of knowledge continues to rise (Figure 1-1). This reaction to a little knowledge about biosolids was confirmed by the survey conducted for this project, which found that people have some initial uncertainty and quickly come up with relevant concerns and questions.

This model has important implications for biosolids managers. If people learning about biosolids recycling only have time and energy to get as far as time "A" in Figure 1-1, they will know something about biosolids recycling, but will likely remain uneasy about it—especially if the initial information they have received raises questions. If the biosolids manager is put off by the lack of acceptance and abandons the project, unwilling to work through the conflict, he or she leaves the situation at time A—at the lowest point of acceptance, when the public's knowledge is still minimal. The result is a lasting negative impression and unresolved questions.

On the other hand, if the biosolids manager encourages people to take the time and continue the process of learning about biosolids management—thus acquiring more knowledge and getting to time B or beyond—questions may be answered and the uncertainty may be reduced, resulting in increased acceptance. This is the experience in some situations, as reported by Logan and others.

Some social scientists might argue, however, that public acceptance is created not by increased knowledge as much as by the act of listening and caring enough to respond respectfully. That is, the process of communication and developing the public relationship is more important than the actual knowledge. This perspective is reflected in the following:

Model 2. Conflict is triggered by a lack of shared knowledge about biosolids–as well as by emotional, social, cultural, and value differences. When there is interaction and a sharing of information, mutual understanding improves, agreement is reached, and the conflict is reduced.

With emphasis on *dialogue*—the mutual sharing of information and understanding—a biosolids/sewage sludge conflict may mature over time from widely divergent views and positions toward greater mutual understanding and a smaller gap between positions (Figure 1-2).

Figure 1-2. Another Possibility: Divergent Views Move Over Time Toward Mutual Understanding.

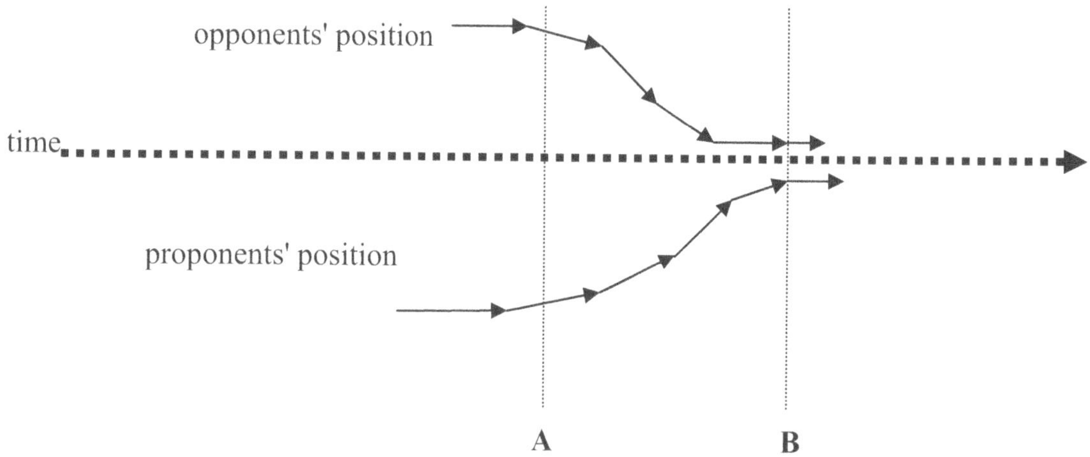

Once again, if the development of the debate is arrested too early (i.e., near time A in Figure 1-2), the widely diverging perspectives and positions in the debate will be locked in place. This scenario may result, for instance, when a biosolids manager decides to avoid conflict and abandon a project at the first hint of public concern. In contrast, if the local policy debate is allowed to mature over time, it is possible (though not guaranteed) that time B can be reached and the gap can be closed—at least to a large extent.

In both of the above models of conflict, it benefits the biosolids manager to work through the process and try to get at least as far as time B. The result will likely be a more mature discussion focused on more substantive issues and the "grays" in the middle of the debate. This can be a productive dialogue in which each side is learning from the other and might lead to a better outcome, including a more environmentally beneficial biosolids management program. Such dialogue may also improve long-term relationships.

Model 3. Conflict is triggered more by emotional, social, cultural, and value differences–and lack of shared information exists, but is less important. When there is interaction and a sharing of information, the value differences only become clearer and the conflict becomes entrenched and may increase.

If the individuals involved in the initial interaction are incompatible—e.g., have differing values or communication styles—then it is possible that no rapport develops and, on the contrary, there is wariness and/or growing distrust. Thus, as time B is reached, an entrenched conflict is developing and the positions held by the two opposing sides may actually move farther apart (Figure 1-3).

Figure 1-3. A Third Possibility: Entrenched Conflict Develops Over Time.

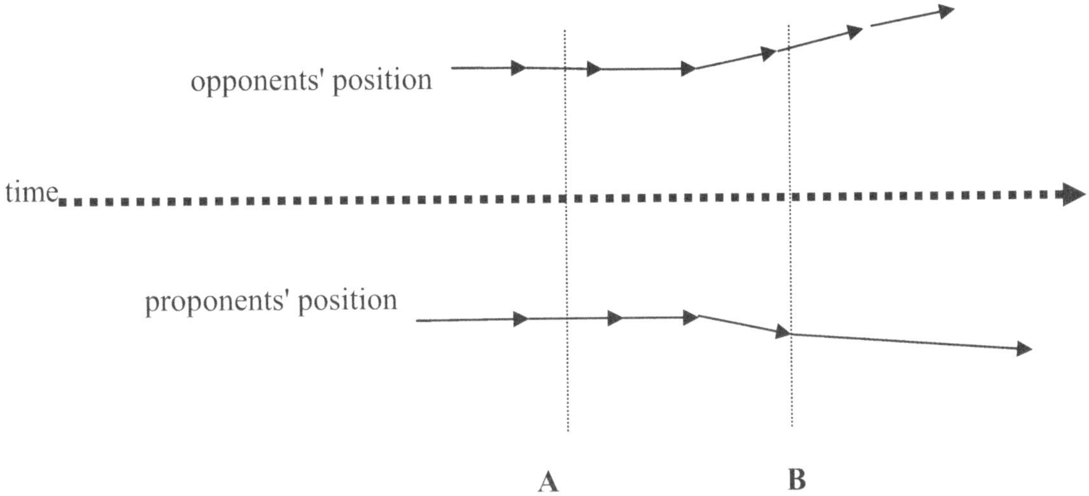

This scenario has developed in many biosolids recycling situations (e.g., Chapter 2.2.13). Conflict of this sort is described by Birkhoff (Water Environment Research Foundation, 2001), who summarized some of the social science literature:

> [W]ithout productive engagement, disagreements over policy choices may escalate to antagonism between parties. These hostile interpersonal dynamics develop through psychological and social psychological processes. They may begin instrumentally, as one party seeks to differentiate itself and its view on the issues from the other parties…. The community also goes through a predictable cycle as the conflict escalates. Opposing side[s] may become increasingly polarized. New organizations and leaders may spring up to further different interests. Existing community organizations are drawn into the conflict and may increase polarized relationships. Finally, word of mouth communication increases as people begin to mistrust traditional sources of information and news.

It is important to recognize entrenched conflict as soon as possible, in order to discontinue the forms of communications and actions that tend to further increase the conflict. Once conflict is entrenched, it is usually necessary to apply conflict management and conflict resolution strategies (for more information, see Sandman, 2000; Susskind and Field, 1996).

As you develop a public participation process for your biosolids recycling program, think about any conflict you are in and what you can do to help reduce it—but don't be afraid of conflict or run away from it. Be willing to work through it in a respectful, open, honest, and caring manner. Conflict, as Birkhoff and others point out, is an integral part of social interactions in a democratic society. There are many examples of sound decisions coming from public processes that involved considerable conflict. For example, a Vermont town on the shore of Lake Champlain recently spent several years struggling with developing a plan that was acceptable to all parties for protecting the local lake water quality. An initial public citizens advisory group failed to produce an acceptable plan. However, because of the learning and public involvement that occurred during that process, a second advisory group was able to efficiently develop a plan that was accepted by all.

1.5.4 Determine What is Beyond Your Control; Work On What You Can Control

There are many factors that affect the public perceptions and acceptance of a biosolids program. Many of these factors directly affect the development of public relationships. For example, if a neighbor to a proposed land application site has had a negative experience with the farmer proposing to use biosolids, that former experience will have a significant impact on the relationship between the biosolids manager and the neighbor and, thus, on the proposed program.

Biosolids managers often note (and complain about) such factors, and some are discussed in the case studies (see Chapter 3.0). However, the biosolids manager needs to recognize which factors are generally beyond his or her control (Table 1-3). Pre-existing experiences, predilections, and prejudices cannot always be predicted. And they can only be avoided if the choice is made to refrain from pursuing any communication or relationship whatsoever—which means abandoning the project. While it is distracting and unproductive to focus much attention on uncontrollable factors, the biosolids program manager still needs to know about them and understand their potential influence on the biosolids program. The best way to learn about and understand these factors is to develop relationships with stakeholders and keep the lines of communication open.

If you find yourself complaining about something, stop and think about whether it is something you can control or not, or if you can avoid it, and determine its possible impacts. Then move on.

It is also possible to use this kind of analysis of factors to aid in choosing between alternative biosolids management options. For example, if a pre-existing negative relationship between a neighbor and a farmer is likely to translate into opposition for biosolids use, the biosolids manager may decide to choose a different location, make additional concessions, or select another biosolids management technology.

Table 1-3. Some Factors Out of Biosolids Managers' Control.

FACTOR	POTENTIAL IMPACTS	CASE STUDY EXAMPLES
Pre-existing relationships between local landowner (farmer or facility site) and neighbors	Communities often split into opposing groups along existing "fault" lines; a biosolids program could cause a wound to re-open.	New Hampshire (Chapter 2.2.14) Denver's Metrogro Farm (Chapter 2.2.9)
Urban/rural conflict; perceived or true imbalance of power between urban and rural communities	The biosolids manager may be starting with a strike against her or him. It will be an added hurdle to establishing trust and a sense of fairness.	New York City and Rural Communities (Chapter 2.2.8) Philadelphia Biosolids Mine Reclamation Program (Chapter 2.2.10) Chicago, IL (Chapter 2.2.5)
Changes in land use near application sites	Varied expectations re quality of life, community	Boulder, CO (Chapter 2.2.6) Montgomery County (Chapter 2.2.11)
Exaggerations and misinformation, people become concerned because of what's on the Internet	Means more work to present useful, accurate information and to have dialogue with stakeholders.	New Hampshire (Chapter 2.2.14) Philadelphia Biosolids Mine Reclamation Program (Chapter 2.2.10)
Scientific dispute; uncertainty	People are often concerned when they become aware of uncertainty in the scientific basis of what to them is a "new" technology. Risk communications may be useful.	New Hampshire (Chapter 2.2.14)
Past issues sensitize the community.	What's happened before cannot be changed, but will influence how people react to new conflicts.	New York City and Rural Communities (Sierra Blanca, Chapter 2.2.8) Denver's Metrogro Farm (Chapter 2.2.9)
Regional social differences (e.g. public familiarity with farming, denser or less dense populations, history of public acceptance or not, level of environmental interest and concern, etc.)	Some existing factors aid a biosolids manager, such as familiarity with organic waste management. Information provided should be appropriate for the audience.	New York City and Rural Communities (Prowers County, Chapter 2.2.8) New Hampshire (Chapter 2.2.14)
A new thing - introducing new technology	People are leery of what they perceive as "new" technology.	King County, WA (Chapter 2.2.1) Everett, WA (Chapter 2.2.12)
Distrust of the biosolids manager who has a stake or a financial interest.	Will require more effort to overcome distrust.	Kern County, CA (Chapter 2.2.13)

1.5.5 Determine Who Your Stakeholders Are

* a *most critical* action step

Who may potentially be impacted by your program? Imagine how people will learn about your program. Will their experience of learning about it likely be a positive or a negative experience? Table 1-4 is a list of possible stakeholders and some of the potential impacts and concerns that may affect them. Not all of the impacts listed will be significant in every project; it may be useful to further develop this table for your particular program.

Table 1-4. Identifying Biosolids Management Program Stakeholders.

Stakeholder	Impact: Perceived Benefit	Impact: Perceived Risk	Degree of Impact (as perceived by that stakeholder)	Other information about stakeholder
Farmer/biosolids user	Fertilizer, cost savings	Liability, trace contaminants, nutrient control, pathogens	Benefit: great Risk: some to minimal	May have had plenty of time to study biosolids.
Biosolids land application workers	Employment	Trace contaminants, pathogens, heavy machinery	Benefit: great Risk: minimal to moderate	Familiar with biosolids—experienced.
Neighbors to utilization site or facility	Possible appreciation that farm/open space remain viable	Odors, air emissions, pathogens, lower property values	Benefit: minimal to none Risk: minimal to considerable	Unfamiliar with biosolids, concerned about something new.
Local officials in the area of the biosolids land application use	Commercial activity, economic benefit to farms, landowners	Liability, safety, citizen concerns	Benefit: minimal to considerable Risk: minimal to considerable	May not be knowledgeable about biosolids; motivated by what concerns citizens.
Local citizen or environmental group	Open space or the benefits of cleaning water that leads to sewage sludge production	Citizen concerns, potential for pollution, mismanagement, public health, long-term impacts	Benefit: minimal to none Risk: potentially considerable	Watch dogs and instigators of change, sometimes an important societal function.
News media	Possibly interesting story; may be required coverage since it deals with public works	Difficult story to write in a balanced way, with technical accuracy. Complex. People are upset.	Benefit: minimal Risk: minimum to considerable (discomfort due to conflict, complexity)	May or may not pay much attention or spend time learning; local press can have considerable influence.
Wastewater facility/generator (and the local officials who oversee the utility)	Biosolids use, cost efficiency (political standing)	Liability, proper management, public image (political standing)	Benefit: considerable to huge Risk: minimal to considerable	This stakeholder's interests are the major driver in most biosolids programs.
Sewer users, septage haulers	Efficient sewer rates	None, unless due to illegal discharge	Benefit: minimal to considerable Risk: none	Both groups often know little about biosolids recycling and the benefits they get from it.
Neighbors to truck route/drivers/road	None	Highway safety, damage to roads,	Benefit: none Risks: minimal to	Can be a big concern if there

crews		noise, odors, dust	considerable	are lots of trucks.
Area farmers/agricultural interests	Cost efficiency for farm, soil improvement	Potential impacts, nutrient management, public perception of regional crop quality	Benefit: minimal Risk: minimal to considerable	Farmers and other providers of food and other consumer products are concerned about perceptions of their products and may choose to avoid being associated with biosolids use if they see the public is concerned.
(who else?)				

1.5.6 Get Input from Key Stakeholders as Soon as Possible, Early in the Process

*** a *most critical* action step**

Your biosolids management program needs to assess the importance of developing relationships with each of the identified stakeholders. You can prioritize who should get more attention, but don't leave out any of the obvious stakeholders as you develop a two-way communications plan. You should keep track of what stakeholders are thinking and how they perceive your program. Develop a system for tracking interactions. Make a special effort to talk to and listen to neighbors, communities where biosolids are managed, and local political leaders. Ask them for assistance in making your program more compatible with its surroundings. Provide stakeholders with opportunities to help plan, implement, and monitor the program. Research clearly shows that involving stakeholders as early as possible in the process engenders greater trust and better long-term relationships. And stakeholders are more be willing to be involved if they know they can make a difference—something that is more likely to happen if they are involved early in the process.

To what extent should you involve stakeholders? You need to involve them enough so you can have confidence that you know what they are thinking and feeling about your program. Some stakeholders and situations will require more extensive public involvement. For example, there may be neighbors who are especially concerned and should be engaged in helping make the program work better in their community. For them, you might establish a Citizens Advisory Committee and a complaint hotline. In some situations, state and federal oversight is perceived as minimal, so the biosolids manager may want to encourage and support community efforts to provide local independent monitoring that will help build trust and confidence.

Throughout the process of developing public outreach and participation, the biosolids manager will need to assess how extensive the public participation process should be. Susskind (personal communication) delineates four levels of public involvement, from some to a great deal:

- write a fact sheet, have a tour;
- develop multi-stakeholder dialogue, build relationships;
- develop joint fact-finding and collaborative activity; and
- commit to implementing programs together.

How far you want to go with stakeholder/public participation is up to you and your stakeholders. In general, it is better to offer the public *more* responsibility and *more* opportunities for input rather than too little. (Even if they don't take you up on the offer, they appreciate it and will trust you more.) If they get involved, you have to make it worthwhile for them. Avoid token gestures toward public participation—they can be worse for your program's image than not involving the public at all. In this report, the authors focus on the second and third items in the above list: dialogue, building relationships, and developing joint fact-finding and other collaborative efforts around issues of concern.

While pursuing stakeholder involvement, the biosolids manager should not lose sight of his or her ultimate job responsibility for managing the biosolids in accordance with legal and agency requirements. The biosolids manager still will be held accountable and his or her expertise must be utilized throughout the process. Involving stakeholders does not mean giving up expertise, responsibility, or all control. Nor does it mean giving up robust science and engineering in your biosolids management program. And it does not mean that there will be agreement about all issues, or that everyone will become friends.

Finally, what about those "stakeholders" who inevitably show up to disrupt the process, determined to stop your biosolids management process? Expect them and plan how to deal with them. It is good to involve these people, to some extent, and listen carefully to their concerns and take actions to address those that make sense. At the same time, it becomes even more critical to work with a diversity of other stakeholders in order to obtain a broader view of public perspectives and craft a biosolids management program that works for the largest possible percentage of the community. That is the biosolids manager's responsibility.

Ideally, extreme points of view will marginalize themselves—and the biosolids manager can just graciously stand back and let them do so. However, sometimes it will be necessary to efficiently, but respectfully, marginalize extreme points of view and move on with constructive dialogue amongst the other stakeholders (who will, ideally, include someone who can constructively represent some of the perspectives of the marginalized person(s)). This is hard to do unless you have developed relationships from working with the broad diversity of stakeholders who will support you through this process. It should be done only with care and the input of some of those other stakeholders. This is one reason why you do not want to put your efforts into working only with dedicated opponents—you will need others to work with you if difficulties arise.

1.5.7 Redefine Success

It is important to understand what constitutes "success" and what constitutes "failure" of a biosolids recycling program. Some in the business of recycling biosolids might suggest that a program has achieved success if it results in biosolids hitting the ground. Some opponents to biosolids recycling would consider it a success to stop any biosolids from hitting the ground.

This report's authors recommend a more exacting definition of success that includes four elements, where each is a building block that further strengthens a program's potential for long-term success:

- Biosolids are managed in a technically sound program in accordance with applicable laws and best management practices…
- With the awareness, understanding, and a reasonable level of input on the part of informed neighbors, local communities, and other interested stakeholders…
- With the acceptance, or support, of many stakeholders (they are satisfied, have trust, and feel they are treated fairly and with respect)….
- Consistently, over time, without dramatic fluctuations in levels of conflict or public trust and support.

Each of these components of success is discussed below.

1.5.7.1 Biosolids are Managed in a Technically Sound Program in Accordance with Applicable Laws and Best Management Practices…

A "successful" biosolids management program should not only comply with applicable laws, but also should adhere to best management practices. By going beyond compliance, this definition "sets the bar higher" than is currently the case in some biosolids management programs. The federal Part 503 rule focused on issues related to biosolids quality, application rates, and significant potential impacts; with a few exceptions (e.g., setback distance from surface waters), Part 503 does not stipulate management practices. A growing number of states, however, now include best management practices in their biosolids management regulations. Overtly and formally including current best management practices in a biosolids recycling program can help move the program toward the new definition of success.

In the experiences noted in this report, if a program is not in compliance with legal requirements and best management practices, it will eventually run into public opposition/perception issues. An illegal program may be shut down by regulators. But beyond issues of legality, if people who are initially uneasy about biosolids recycling discover that the program near them is not being run in accordance with best current practices, they will (often successfully) use that fact as a rationale for stopping or changing the program.

Programs that are perceived as poorly run create public distrust and even outrage, and their managers will lack credibility. The fact is that all stable, successful biosolids recycling programs today go above and beyond regulatory requirements. Many have negotiated with stakeholders to establish special practices that meet the needs of the local situation.

1.5.7.2 …With the Awareness, Understanding, and a Reasonable Level of Input on the Part of Informed Neighbors, Local Communities, and Other Interested Stakeholders…

There are many currently operating biosolids recycling programs that appear to be operating smoothly in the absence of informed neighbors. No one is complaining, and the biosolids are going on the ground.

However, as the experiences in the literature review and case studies show (see Chapters 2.0 and 3.0 on the accompanying CD-ROM, biosolids managers in such situations can be surprised by sudden public opposition. Such programs are vulnerable because the biosolids managers have no information on what stakeholders are thinking or how stakeholders may react.

Meeting this second requirement for success is probably the most difficult hurdle for the biosolids manager. The recommendation is that existing biosolids programs stand up and say "we're here, and this is what we're doing." This first proactive step toward fully informing neighbors and other stakeholders about a biosolids recycling program can be a frightening experience.

In some instances, this step creates an explosion of concern, anger, and turmoil. This is often referred to as "outrage," which Peter Sandman (2000) defines as "everything else that the public considers part of risk" besides the actual calculated death rate hazard (e.g., one death in a million). His theory is that the public perceives risk as a sum of hazard and outrage. Thus, the minimal risk that some scientists have calculated (and that biosolids managers actually experience as they work) will be increased, in the minds of the public, by the addition of outrage factors (Table 1-5). The more of these factors operating in a particular biosolids recycling program, the greater the perceived risk of the program.

Table 1-5. Outrage Factors that May Increase Public Perception of Risk (after Sandman, 2000).

OUTRAGE FACTOR	EXPLANATION
Involuntary	Imposed on the public stakeholder, not something he or she chose to have happen
Industrial	Not natural.
Uncontrollable	Out of personal control. For example, we accept the high risk of driving because we feel we have a lot of control in the situation.
Exotic	Not familiar, strange. For example, we accept the high risk of driving in part because it is familiar.
Unknowable	Some uncertainty exists; this is common in science-intensive topics, where scientists cannot and will not say that something (like biosolids) is 100% safe. Similarly, the exact content of any particular bag or load of biosolids is somewhat unknowable.
Unfair	For example, a neighbor may feel it is unfair to have to put up with odors when he or she receives no apparent benefit from a biosolids program.
Memorable	Odors make strong, memorable impressions on the brain, and this can be significant for biosolids programs. In addition, other "yuck" factors can make biosolids recycling negatively memorable.
Concentrated in time and space	For example, a Class B biosolids land application event, with associated odors, tends to be concentrated in time and space.
Dreaded	Feared. For example, with a biosolids program, a neighbor might be scared thinking about what could happen to his or her well water or health.
Morally relevant	Moral arguments can be made about it. For example: Is it moral and fair for urban wastes to be recycled on rural land?
Created by a process that is seen as closed	For example, a local community can experience the permitting process as closed if they are not invited to be involved much.
Untrustworthy	Outrage is increased if those proposing the program are seen as untrustworthy or if the system cannot be trusted.
No perceived benefits	For example, many biosolids programs provide no obvious benefits to the neighbor or other local stakeholder.

The survey conducted for this project (see Chapter 4.0) corroborated the significance of several of these factors. For example, the survey respondents' concern was increased when industrial sources were mentioned.

Therefore, the process of standing up and saying "we're here" needs to be done with care—and with the help of risk communications (see below). By reducing the number of outrage factors—for example, by inviting stakeholders to provide input—you can reduce perceived risk. For example, many Class A biosolids recycling programs (e.g., Tagro in Tacoma, WA and biosolids composts in San Antonio, TX and Merrimack, NH) have developed public support by working with communities and customers so that the direct benefits are clear to homeowners and other citizens through demonstrations and word-of-mouth recommendations from satisfied customers.

When you stand up and say "here we are," you are not actually creating negative public reactions; you are discovering public concerns that were unknown to you. By instigating the process, the biosolids manager can be prepared for the ensuing public discussions and education process and will likely be credited for his or her outreach effort and openness. It is far better to be at least partially in control of this public dialogue process rather than be caught by surprise when your program is suddenly "discovered" by a negative media storm Also, remember that working through discussions and conflict can help build long-term public relationships—which should remain the focus of any biosolids public outreach effort.

Finally, biosolids program managers are not the only ones who must be involved and prepared for this critical second step toward a higher standard of success. Those with whom the biosolids program partners—farmers and customers—need to be informed and involved in the risk communications process, especially if they are going to be caught up in the discussion or conflict. For some biosolids managers, a significant hurdle may be convincing a farmer to work together on being more open and communicative with neighbors and the surrounding community.

1.5.7.3 ...With the Acceptance, or Support, of Many Stakeholders (They are Satisfied, Have Trust, and Feel They are Treated Fairly and with Respect)...

If a biosolids management program is going to avoid costly conflict, it will need to have some community *support*. Once dialogue with key stakeholders, including neighbors, has been established (see more on developing public relationships, below), it is likely that many will accept or openly support the program. This has been the experience of many programs.

The biosolids survey (Chapter 4.0) confirms that most people don't know about biosolids and that when they first learn about them, many have questions and concerns. However, such people find arguments that biosolids recycling "disposes of a waste product" and "returns nutrients and organic matter back into the soil" to be effective and reassuring. Other survey results indicated what the public tends to be concerned about and what can reduce their concerns. These results point to actions that biosolids managers can take to reduce public worry about biosolids:

- When people have questions, most ask their neighbors and then turn to state, federal, and academic experts for answers. If these experts are informed about your program, they will be able to answer questions with confidence and objectivity, which reduces public concern.

- People favor constructive uses of biosolids (e.g., extracting energy from biosolids or recycling nutrients), so biosolids recycling programs should be able to clearly demonstrate such uses.
- People are concerned about contaminants and pathogens in biosolids, and these issues, as well as source reduction and pollution prevention, should be addressed and discussed forthrightly.
- People are concerned about industrial waste in biosolids. Pretreatment programs should be robust and the public can be invited to learn more and help with this crucial aspect of creating high quality biosolids.
- People are very supportive of wastewater treatment and, therefore, should understand the necessity of managing biosolids.

This third step in the definition of success does not suggest that *every* stakeholder has to support the biosolids management program for it to be a success. There will always be people who disagree with the concept and/or the way it is being implemented. The goal of this element of success is to ensure, through an educational process and dialogue with the public, that there are at least a good number of stakeholders, representing differing perspectives, who accept your biosolids management program. Over time, the number of supporters may well increase if the program is well run and people witness its benefits and minimal risks.

1.5.7.4 …Consistently, Over Time, Without Dramatic Fluctuations in Levels of Conflict or Public Trust and Support.

Many biosolids recycling programs are able to manage several years of relatively good "public acceptance." However, true success comes when the first three steps above are in place, *and* the level of public support does not suffer dramatic swings over time. In other words, even if biosolids are accidentally spilled on a highway or there is a national news story about problems at some other biosolids recycling program, there are enough informed local stakeholders and citizens to work together to address the concern in a cooperative, positive, problem-solving way, rather then reacting with outrage.

As Figure 1-4 illustrates, building a biosolids management program that is truly "successful" over the long term is a step-by-step process involving the public. The steps outlined above serve to build trust and relationships. Even when challenges arise, a biosolids recycling program built openly around relationships—with neighbors, elected officials, regulators, contractors, treatment plant personnel, environmental and public health groups and other stakeholders—is equipped to handle a situation, calmly solve a problem, and move on.

Figure 1-4. Steps Toward a New Definition of "Success" for Biosolids Recycling Programs.

1.5.8 Build Public Relationships

*** a *most critical* action step**

1.5.8.1 The Common Denominator of Success: Building Relationships

One goal of this WERF research project has been to identify what actions are consistently effective in developing positive public perceptions and public support for biosolids management. A theme that runs through many programs that have gained public support is *building public relationships*.

Relationships are what many of the managers of successful long-term biosolids programs talk about. Almost all of the case studies (see Chapter 3.0 on accompanying CD-ROM) include some discussion of how relationships had—or would have had—a significant impact on public perceptions of biosolids management. Biosolids programs that have gained public acceptance have developed relationships with critical stakeholders—neighbors, agricultural advisors, researchers, government officials, the media, concerned citizens.

Lawrence Susskind and Patrick Field (1996) begin their book, *Dealing with an Angry Public*, with this: "There are many reasons for the public to be angry. Business and government leaders have covered up mistakes, concealed evidence of potential risks, made misleading statements, and often lied....Wouldn't you be angry if you had been hurt, misled, or threatened?" Biosolids managers should recognize that the public's reaction to them will be colored by past experiences with other public officials, scientists, or engineers who may have misled or lied to them. The way to overcome this possibility is to develop a relationship with that citizen. Only through beginning of dialogue, through building a relationship, can a biosolids manager begin to gain the confidence of citizens in a community.

Many successful biosolids managers over the years have noted the importance of building relationships with "the public," with "stakeholders," with "gatekeepers":

- In Charlotte, NC, biosolids manager Trille Mendenhall, an active Sierra Club member, developed relationships with other environmentalists who learned about biosolids recycling from her.
- In King County, WA, former biosolids manager Peter Machno developed relationships with local farmers and, in the Mountains-to-Sound Greenway project, with a variety of environmental leaders (see Chapter 3.2.1).
- In the City of Wichita, KS, the Biosolids Management Plan for land application "outlined and pursued a public education and acceptance program for the affected agricultural community" by establishing "credibility and an understanding working relationship" (Vaughan et al., 1994).
- In Austin, TX, relationships developed between the wastewater treatment facility and the local Audubon society and its birdwatching members because storage ponds provide precious habitat for birds.

1.5.8.2 Attributes That Earn Trust and Build Relationships

There are specific personal and organizational attributes—ways of acting and communicating—that help earn trust and build public relationships. Many of these attributes (such as honesty, fairness, and following through on commitments) are the same kinds of attributes that build personal relationships. Table 1-6 lists some key attributes, along with references to case studies in Chapter 3.0 (on the accompanying CD-ROM). Consultants and academic specialists in the fields of public participation and communication can assist biosolids managers to incorporate these attributes into their programs and their personal interactions with the public.

Table 1-6. Some Key Attributes for Building Relationships.

ATTRIBUTE	CASE STUDY REFERENCE	MORE INFORMATION
Personal and Organizational Attributes		
Working to improve, with public input, environmental goals and outcomes; Environmental Management System (EMS)	3.2.9 Denver's Metrogro Farm	National Biosolids Partnership, www.biosolids.org
Openness, transparency, sharing information	3.2.1 King County, WA 3.2.6 Boulder, CO; 3.2.8 New York City and Rural Communities (Prowers County)	Sandman, 2000 Susskind and Field, 1996
Honesty–admitting mistakes, humility, not overstating accomplishments	3.2.6 Boulder, CO, 3.2.8 New York City and Rural Communities (Prowers County)	Susskind and Field, 1996
Fairness, respect, equity, sharing power	3.2.6 Boulder, CO 3.2.7 New York's CAC	Hartley, 2003 Renn et al., 1995
Excitement and belief in what you're doing, being personable	3.2.1 King County, WA 3.2.2 Milwaukee, WI 3.2.5 Chicago, IL 3.2.8 New York City and Rural Communities (Prowers County)	Toffey, 2000(d)
Compassion		Sandman, 2000
Being predictable, meeting expectations, doing what you said you would do, integrity, trust	3.2.12 Everett, WA	Susskind and Field, 1996
Ensuring adequate resources to do the job right (reducing costs, meeting budget, and profit are not the only things that count).	3.2.1 King County, WA	
Supporting consistent staff and procedures to help support long-term relationships	3.2.1 King County, WA 3.2.9 Denver's Metrogro Farm	
Communications & Information		
Ability to listen and hear, allowing others to	3.2.1 King County, WA	

talk, empowering others.	3.2.2 Milwaukee, WI 3.2.12 Everett, WA	
Acknowledgement of others' concerns, being responsive.	3.2.6 Boulder, CO	
Clarity, using appropriate modes of communication (e.g. avoiding technical jargon), communicating technical information in ways the audience can understand but is not patronizing or overly simplistic.	3.2.1 King County 3.2.6 Boulder, CO 3.2.8 New York City and Rural Communities 3.2.12 Everett, WA	Sandman, 2000 Wantland, 2002
Knowing risk communication concepts.		Sandman 1987(a & b)
Follow-through; providing answers.	3.2.12 Everett, WA	
Having support and communication via a variety of people; third-party support and spokespeople.	3.2.1 King County 3.2.8 New York City and Rural Communities (Prowers County)	
Communication by the ultimate responsible party (e.g. the public utility biosolids manager rather than just the contractor), even in distant biosolids use localities.	3.2.1 King County, WA 3.2.10 Philadelphia Mine Reclamation Program	Susskind & Field, 1996
Reducing surprise (telling people in advance). Providing information to media and public proactively, communicating benefits and risks forthrightly and honestly.	3.2.1 King County	Wantland, 2002; Rocky Mountain WEA
Having the right person communicate; training for biosolids program communicators.		Wantland, 2002; Rocky Mountain WEA
Knowing needs of the other (e.g. knowing media deadlines)	3.2.14 New Hampshire	Rocky Mountain WEA; Wantland, 2002

1.5.8.3 Choosing the Right Tools

"Tools developed jointly are relationship-building." --Lawrence Susskind

After reviewing many different experiences, the authors conclude that the most useful, efficient tools for developing public outreach and participation are those that further develop relationships with stakeholders. Such tools also are most likely to improve perceptions and support for the biosolids management program. (This assumes, of course, that the program is in compliance and follows best management practices, so that the public outreach program is built on integrity and excellence.)

Here is an example. If a biosolids manager has determined that the neighbors to a land application site don't know much about the program, she has many choices of how to inform them. She could send out a fact sheet, write a letter to the editor, or get the paper to write a news story. All of these actions would provide some one-way communication of information, but would not be likely to further build relationships. Holding an informational meeting (hearing) at which technical experts provide detailed technical information about the benefits and risks may involve some dialogue that develops relationships with the public a bit more—but the technical presentations are mostly one-way communications and can result in distrust and resentment.

Better choices for the biosolids manager would be attending neighborhood meetings to informally discuss what is going on and ask if people have any questions or concerns. Or the biosolids manager can invite the neighborhood to a tour of the wastewater treatment facility and farm so they can see first hand how the program operates and what steps are taken to address concerns. Or the biosolids manager could invite community members to participate in a multi-stakeholder committee assisting the agency in developing new odor control measures. These options are recommended because they involve opportunities for building relationships.

Table 1-7 includes a list of common tools for developing public outreach and communication programs. You may think of additional tools that you can add to this list. For tools that you are considering using, answer each of the questions in the matrix. Sometimes, the same tool used in different situations will result in different answers to the questions. For example, while a letter to the editor is often a simple one-way communication that does not build dialogue, in some instances—such as when it is in response to a request for specific information and its appearance will show responsiveness and enhance trust—it is an appropriate part of dialogue and advances relationship building.

Table 1-7. Evaluating Public Outreach and Participation Tools.

TOOL	WHO?	COMMUNICATION & INFORMATION					RELATIONSHIP BUILDING				COST
	Which stake-holder(s) does it reach?	Does it educate stakeholders?	Does it improve understanding among stakeholders?	Does it improve understanding among biosolids staff/mgmt?	Does it increase credibility of information?	Does it provide an on-going means of communication?	Does it provide an on-going means of problem solving?	Does it help resolve conflict?	Does it share decision-making?	Does it improve relation-ships?	Cost?
PRINT/PHONE/MEDIA											
Phone/ email complaint hotline											
Consumer surveys											
Fact sheets & brochures											
Advertising, product promotions											
Website, email											
White paper											
Newsletters, annual report, updates (contacting before spreading)											
Letters to editors											
News coverage											
Provide B-roll video to TV stations											
Talk radio or television											
(other?)											

PUBLIC MEETINGS/GROUPS											
Facility or site tours, picnics											
Hosting public meetings											
Speaking at community /neighborhood meetings											
Citizen Advisory Committee											
Joint fact finding, co-learning											
Having a local excited and expressive "champion"											
Developing mutually-beneficial projects in cooperation with garden clubs, citizen groups, watershed groups, etc.											
Stakeholder involvement in developing group meetings and process											
Formal conflict resolution process, consensus building											
(other?)											

ASSURANCE MECHANISMS											
Research demonstration sites											
Complaint response team											
Local oversight by local people											
Biosolids applier certification											
Bonding (addressing liability)											
Environmental Management Systems, independent audit											
Contingency funds											
(other?)											
MESSAGES/ KEY TOPICS											
Biosolids recycling puts to use a waste product (that results from the important process of cleaning water).											
Recycles nutrients											

and organic matter to improve soils.											
Biosolids is a natural fertilizer.											
(other?)											

1.5.9 Study Communications, Especially Listening

* a *most critical* action step

Bill Toffey, manager of the biosolids program for the Philadelphia Water Department, noted (2000a): "We are at risk of losing our programs and increasing costs of our programs because of our failure to communicate the benefits of biosolids."

Many public agencies have difficulty communicating the importance of their missions and work. But agencies with primarily technical staff, such as public works departments, often have an added lack of communications skills. Engineers and wastewater treatment facility operators tend to be less skilled at communications, as a group, than, for example, politicians. Their jobs usually don't require any special training in communications. In addition, this field was traditionally male-dominated—and communications research has shown that, through socialization, women are generally better trained in communications than men and are more focused on relationship-building (Tannen, 1990).

Whatever the reasons for the lack of strong communications regarding biosolids recycling programs, it is clear that things are changing—forced, in part, by conflicts. Many larger agencies now have communications specialists and/or are providing communications training for staff. Training organizations, such as the Water Environment Federation and regional biosolids associations, occasionally offer workshops in communications, media relations, and associated topics. Biosolids managers are now encouraged to ensure that they, or someone on their staff, are good communicators.

1.5.9.1 How You Communicate is Important

The social science field of communications is broad. What is most important for biosolids managers to pay attention to and incorporate into their communications? There are certain concepts or schools of thought worth focusing on:

- Listening is as important as talking; biosolids managers should pay attention to what the full variety of stakeholders, including critics, have to say. Listen for the emotional content. Listening with empathy and caring is important.
- "Risk communications" is a specialized field that helps technical people better understand how others perceive and understand risk; biosolids managers can benefit from studying risk communication (e.g., see Sandman; Covello; or Powell, 1996).
- Communicating in times of crisis ("crisis communications") is another specialized area in the study of communications; larger, more complex agencies are well-served to have someone trained in how to manage communications with integrity during a crisis (Hyde, 2002).
- Communication involves much more than the information being presented. People respond to particular attributes or styles of communication, and it is helpful for the biosolids manager to know these (Covello and Wolf, 2003; Wantland, 2002). For example, conveying empathy can have as much impact on gaining trust as all other factors combined.
- Effective communications are beneficial for all stakeholders—including the public—because it ensures excellent exchange of relevant information. So long as your communications are sincere and two-way (mutually beneficial), efforts to make them more effective should not be interpreted as "spin."

- Effective communications involve use of ordinary language, not jargon. Example: don't use "Part 503" to identify the federal regulations, use "federal regulations."
- Effective communications involve many non-verbal cues. Example: When the communicator stands above (on a stage) and distant, he or she is perceived as more intimidating. Instead, arrange a meeting so you can stand closer and at the same level. Example: What a person is wearing can make a difference—less formal clothing will make people more comfortable than very formal attire. Example: How a person responds to questions makes a difference: interrupting or making gestures or faces that suggest "that's a stupid question" is obviously not helpful.
- Communications intended to manipulate or deceive are quickly discerned by the public. Don't do it: it will damage relationships by creating greater distrust and outrage.

1.5.9.2 Apply Survey Findings

The 2002 Biosolids Public Knowledge and Perception Survey (see Chapter 4.0) suggests several things that biosolids managers can do to improve their communications efforts:

- The fact that few people know anything about biosolids/sewage sludge recycling means that there is a need for clear, introductory information.
- Most people see wastewater treatment as a good thing, so discussing how biosolids relate to this important environmental program should help engender positive responses.
- Many people know what kinds of concerns might be involved in biosolids recycling—so effective communications will require expecting, hearing, and addressing such concerns.
- People have established ways of learning about the world, and the biosolids manager's communications efforts will be more effective if they tap into people's existing communications systems (for example, bring your information to existing neighborhood meetings or into existing email networks).
- Telling people in advance of a biosolids land application event and providing important details about what will be happening reduces their level of concern.
- Communications that include opportunities for questions and dialogue with public stakeholders are most effective when done early in the process—and regularly.
- Certain arguments are more effective than others when introducing biosolids recycling. Messages that result in an increased positive response to biosolids recycling are: "biosolids recycling returns nutrients to soils" and "biosolids recycling puts to use a byproduct of necessary sewage treatment."
- Who speaks for a particular biosolids management program is critical. The survey corroborated the fact that some categories of people (academics and sometimes government officials) are more trusted than others. Less trusted are those with a profit motive and wastewater treatment operators who appear associated with the particular biosolids program. Be aware of this and the traits of a good communicator as you decide who will be communicating for your program.
- Advising neighbors in advance of applying biosolids may reduce their concern and increase their trust of the applicator and the process. Some Class B bulk biosolids land application programs ensure that a knowledgeable and articulate local farmer is a central communicator about the local use of biosolids on his or her land and that of other area farmers.

Communications are especially effective if they are honest and direct and do not avoid sensitive topics. Because the public tends to be uneasy about biosolids recycling, in any public forum where such uneasiness is likely to come up, it is important to openly recognize that discomfort. By doing so, you are seen as more trustworthy. Biosolids managers might say things like:

> *I believe that using biosolids recycles critical nutrients and organic matter back to soils and helps us use a byproduct that is created as part of the process of sewage treatment. I know that the idea of taking sewage sludge and carefully treating it so it can be used beneficially like a fertilizer is a troubling concept for many of us. It raises many questionslike "What about all the things—the chemicals—that go down drains? What happens to them? How can this be safe?" There are many safeguards taken to create and manage biosolids. I certainly believe recycling these materials helps society and the environment and is relatively safe, if done correctly. But others disagree, and you may be unsure. I'm glad to discuss concerns and questions you may have.*

Likewise, if a person or audience is interested in learning more, it is helpful to provide information critical of biosolids recycling, as well as information that supports your position, in order to provide the public a chance to understand the various arguments, pros, and cons, and decide for themselves. For example, the survey results clearly show that if there are any industrial inputs to biosolids, public concern is heightened. Therefore, the biosolids manager might best bring up this concern and address it head on. You gain credibility by openly disclosing the variety of information and perspectives on the topic. You lose credibility if you "gloss over" any concerns and/or appear to be hiding anything.

Finally, public outreach and participation around biosolids management programs will benefit dramatically by communicating with and involving the public as early in the process as possible. This cannot be overemphasized. Social science literature on public participation strongly emphasizes the importance of public outreach and participation early in the process.

Within the biosolids field, the focus groups and survey work conducted for the Greater Vancouver Regional District (2000) clearly found that public respondents felt "they needed more time and more information to form an opinion on biosolids use." Why? After hearing the benefits of biosolids recycling, people need to be given a chance to know and think directly about where biosolids come from, so they can wrestle with their initial uneasy reaction. They need time to think about it and they need opportunities to ask questions and obtain answers and additional information. They need to be able to make up their own mind, based on full, accurate, diverse information.

With some creative, proactive planning, biosolids managers have the ability and authority to create project schedules and activities that ensure that critical stakeholders—including neighbors and community leaders—have the time and support for this process of learning. Too often biosolids managers let regulatory schedules dictate when and how communications happen. Such regulatory frameworks are not developed with sensitivity to public concerns in mind. For example, many state regulations require just a letter of notification to neighbors, followed soon after by a public hearing. The state agency then has a limited time before they have to provide the permit or deny it. This time frame is too short and includes no opportunities for constructive dialogue. The letter of notification is a one-way communication tool (although it often includes a way for people to be in touch with the biosolids manager). Further, the public hearing usually is not conducive to developing strong two-way communications—it's too short and it's seen as a one-chance event at which people feel pressured to make their case strongly (see below). Biosolids managers should develop a communications schedule that ensures effective communications and adequate time for the public to absorb information. Then see how to fit the regulatory requirements into that schedule.

1.5.9.3 The Word "Biosolids"

For two reasons, it is a good idea to use the word "biosolids" to describe those sewage sludge materials that have been treated and tested and can be used as soil amendments or fertilizers.

First, as those who helped create the new word a decade ago suggested (Powell-Tate, 1993), the term "sludge" is inexact and may be confusing; it refers to a wide variety of things, including hazardous wastes and what is found in the bottom of automobile oil pans or glasses of chocolate milk. A precise term for treated and tested sewage sludge is useful.

Second, as determined by this project's survey, use of the term "sludge" creates a measurable significant negative response in the general population in comparison to use of the term "biosolids." The use of "biosolids" seems to be particularly important when introducing the concept to people, as their initial impressions may predispose them to either accept or reject information that is offered later, which, in turn, creates the context within which they will form an opinion about biosolids recycling.

However, the term "biosolids" should not be used as a way of hiding the origin of the material. If today's skeptical public senses secrecy and manipulation in the communications of public officials and other authorities, they can quickly become distrustful and angry (see Chapter 3.2.14). Despite the relative newness of the term, there already are concerned citizens and dedicated opponents who bristle upon hearing "biosolids" because they feel its use is a public relations tactic (Stauber and Rampton, 1995). Early in the discussion, therefore, make sure that your audiences know that "biosolids" refers to a product derived from sewage sludge. Emphasize that the term is used to differentiate between those sewage sludges that meet certain standards and those that do not. Biosolids are treated and tested, can be recycled, and provide benefits to soils and crops.

If, during a public meeting, you encounter suspicion of the term "biosolids," you may want to use the term "treated sewage sludge" at least some of the time. The important thing is not to insist that a particular term be used, but rather to work to make constructive dialogue more likely. Remember to ask yourself whether your language will improve your program's relationships with key stakeholders. If exclusively using the term "biosolids" gets in the way, then don't use it exclusively.

1.5.9.4 Remember Whom You Are Talking To and How You Are Perceived

When providing biosolids information to wider audiences, remember that most people will be hearing about the topic for the first time. This large majority is whom you should be speaking to. They form the mass of undecided observers in the middle of the spectrum of opinion on biosolids (these are the "guardians" mentioned above; at the two ends of the spectrum are the staunch proponents and the staunch opponents).

The undecided majority in the center will begin to form opinions about what you are saying fairly quickly. If you are fair, honest, open, interested in the topic, and have a positive presentation style, you are more likely to gain the trust of this group. If you belittle the opposition, appear arrogant, hedge, hide anything, or have a negative presentation style, you are more likely to make this audience skeptical and they may join the opposition. Remember that they just want to hear straight-forward arguments on the merits of your program. They will make their own judgments.

1.5.9.5 Public Meetings/Hearings

Public meetings are often not conducive to effective communications, especially if there are people present who feel threatened, distrustful, and/or angry. To make public meetings or hearings most effective, advance work and planning are required:

- talk to community leaders and representatives of the various perspectives in advance and help develop together a mutually agreeable format for fair and equitable exchange of information;
- determine clear goals for the meeting (e.g., exchange of information, decision-making) and communicate these to all involved in advance and again at the beginning of the meeting;
- arrange the venue so that people are comfortable (providing a break and food can help);
- don't have "experts" stand up in front, on high, and talk on and on;
- ensure opportunities for all perspectives to be shared;
- visibly record concerns or questions to be addressed (e.g. on a large flip-chart);
- answer questions respectfully, promising answers soon to questions you cannot immediately address (and be sure to follow through on those promises!);
- listen and hear;
- have an independent moderator/facilitator in charge of the meeting's process who ensures fairness and respect amongst all participants.

A particular concern that some biosolids managers have raised is that there often seem to be "hidden agendas" behind many public statements made by concerned citizens at public hearings and in the press. Experts on public perception and participation respond to this concern in this way:

- such concern on the part of biosolids managers are not uncommon or unique to the biosolids field;
- in a public forum—at a hearing or in the press—it is best to *take people at their word* and not try to guess at what "hidden agendas" may lurk in the background;
- don't comment publicly about what you guess may be a "hidden agenda"—you will only incite greater distrust, and, if you are wrong, you will have a public relations disaster on your hands; and
- understand the political landscape, what may be motivating particular people or groups, etc.; this information can help you develop better decisions and strategies for public outreach and participation.

For additional advice on preparing for public meetings and hearings, see, for example, Wantland (2002).

1.5.9.6 Powell -Tate and Beyond

In 1993, the Water Environment Federation contracted with the public relations firm Powell-Tate to develop a communications plan for biosolids recycling. The Powell-Tate communications plan, focused around the message "Biosolids Recycling: Beneficial Technology for a Better Environment", has guided many biosolids managers for the past decade. It has been the only carefully developed strategy for the communications, nationwide, of a concept and group of products that involves millions of tons of marketed materials and billions of dollars of infrastructure and operational costs.

For some biosolids recycling programs, the Powell-Tate communications plan helped immensely by focusing efforts on communicating first with "gatekeepers"—individuals who are respected leaders of public opinion (e.g., see Chapter 3.2.1). Developing long-term relationships with key stakeholders has been critical. Those agencies and biosolids managers that had a naturally effective communications style used the Powell-Tate plan in ways that created better relationships.

The Powell-Tate plan, however, has been attacked as an example of manipulative public relations (Stauber and Rampton, 1995). Its focus on gaining "public acceptance" of the concept of biosolids recycling may no longer be a useful strategy. While blatant marketing of commercial products is routine, the public is less accepting of the concept of biosolids recycling being marketed by public agencies (utilities and regulatory agencies) that are expected to be working for the overall public interest.

Used in the wrong situation, the Powell-Tate plan's emphasis on gaining "public acceptance" can incite anger if it makes citizens feel that they have no input into a decision that may directly affect them. Concerned neighbors of land application sites who feel that their health may have been impacted by biosolids do not welcome a "sell" about the benefits and minimal risks of biosolids. Rather, they require understanding, respect, and a compassionate response to their immediate needs—something that the Powell-Tate communications plan does not address.

Despite its shortcomings, the Powell-Tate communications plan provided useful guidance on how to create message points and communicate effectively. However, biosolids managers should go beyond the "public acceptance" approach and involve the public in meaningful ways, especially around decisions that may affect them. A far better model for public outreach and participation is developing within the National Biosolids Partnership's Environmental Management System (EMS) program (see below).

1.5.10 Provide Information Useful to *Recipients*

Much technical literature on biosolids has been developed over the past decade, ranging from overviews of the regulatory requirements for biosolids recycling to details on agricultural management and trace pollutants. U.S. EPA, state agencies, wastewater treatment facilities, public utilities, trade associations, the National Biosolids Partnership, and biosolids management companies have developed fact sheets, websites, reports, and white papers regarding numerous aspects and issues. Many agencies have also developed effective videos, PowerPoint presentations, and/or websites.

Standard helpful education and outreach materials are:

- A fact sheet for the particular biosolids product, which discusses the production process, testing results, recommendations for use, and where to get more information;
- Highlights of the industrial pretreatment program, household hazardous waste programs, and other efforts that protect biosolids quality;
- A state or regional fact sheet on biosolids recycling, which provides context (e.g., the NBMA biosolids folder and fact sheets); and

- For Class A products, small packaged samples of the product.

While such materials are helpful and necessary, there is much more that can be done. One of the reasons for the common emphasis on developing these traditional written public information tools is the fact that many in the biosolids management field perceive an ever-increasing gap in knowledge between "technical experts" and the general public. However, potential audiences exhibit a wide range of reading skills and technical proficiency. It is key to recognize what kind of audience you are developing materials for. Are you creating something for the average citizen who will be overwhelmed by, and distrustful of, a lot of technical information? Or are you creating something for the concerned citizen who is quite knowledgeable and would be offended by an oversimplification of biosolids issues?

In addition, because the topic of biosolids is less than fascinating to the average citizen, many will just want to know the basics and whether or not it presents any potential significant benefits or risks to them personally. This presents a challenge to the biosolids manager, since much of the biosolids literature is overly complex and technical and will not meet the needs of the average citizen. To make matters worse, when conflicts arise, biosolids proponents often get caught up in a "fact contest," burying opponents with technical information and ignoring where people are and what they bring to the discussion.

Instead, biosolids managers should emphasize listening to and learning from stakeholders and the public before attempting to develop and provide information that will answer their questions and concerns. It is even possible to involve the public in developing the informational materials; e.g., by having a citizens' committee help research, write, and produce fact sheets. People absorb information far more thoroughly if they themselves are helping guide the process of learning.

Example: If presenting information to a watershed group that focuses on protecting water and ecological qualities of a river, consider their perspective as you develop messages for a fact sheet or presentation:

- *Their focus*: Water quality is the number one concern.
- *Your complementary focus*: Wastewater treatment is a key component of maintaining good water quality in the river.
- *Their focus*: Certain land uses create run-off that is non-point source pollution in a river; biosolids use may be a land use of concern.
- *Your complementary focus*: Biosolids are a byproduct of keeping the river clean; biosolids must be managed responsibly.
- *Their focus*: Biosolids recycled in the watershed may impact river water and ecological quality if they run off or leach trace pollutants or nutrients.
- *Your complementary focus*: Proper management of biosolids can address these concerns: agronomic rate, buffers and setbacks, slow-release nutrients in biosolids. Biosolids may help reduce non-point source pollution by reducing erosion, run-off, or leaching because they can stimulate healthier vegetation growth and reduce the need for more soluble chemical fertilizers.
- *Your complementary focus*: Biosolids recycling usually plays a very minor role in a watershed, but biosolids managers can be helpful partners in addressing non-point source pollution because of their understanding of nutrients, pollutants, and agricultural and conservation best management practices.

As decades of education literature has made clear, different people learn in different ways, and information must be available that meets the needs of different learning styles. Most biosolids information currently is in the form of technical written information. This information is helpful to a small percentage of the population who are good readers and analytical, linear thinkers. However, many other learning styles are not served by this, and more needs to be done. Biosolids managers can obtain assistance from graphic artists, educators, or others in developing informational materials for those who learn better through visual images, kinesthetic experiences, and other modes.

Even with improved educational efforts, biosolids recycling will remain of minor interest to most people. Rather than trying to educate the public on biosolids as a separate issue, the biosolids industry should focus on how biosolids fit within larger environmental science topics. For example, information on biosolids recycling could be added to middle and high school environmental science units on wastewater treatment, recycling, and agriculture. A good example of this approach is the Mountains-to-Sounds Greenway Trust's education program, "Seeking Solutions to Sustain Forests," which is part of the curricula in Seattle area schools.

Including biosolids information in a wide variety of informational materials will require collaborative efforts with others whose work and perspectives may be quite different from that of the biosolids manager. However, working collaboratively with watershed groups, environmental educators, conservation districts, schools, or similar groups to bring a mention of biosolids into informational materials accomplishes three goals:

- it provides a chance to discuss biosolids with the group,
- it results in biosolids recycling information becoming embedded in a larger context, and
- it establishes a relationship and dialogue with these important thought leaders.

There is one more consideration of importance to developing informational tools about biosolids: biosolids are complex mixtures and their assessment and quality includes some level of uncertainty. This makes it difficult for biosolids managers to specify the quality of a particular biosolids product with 100% certainty. Biosolids risk assessments are based on assumptions and probabilities that are not always persuasive or comforting to the general public. Thus, the biosolids manager's attempts to discuss scientific uncertainty—using words such as "may" or "approximately" or "minimal risk"—may increase outrage.

One way for biosolids managers to deal with public uneasiness with uncertainty is to engage concerned individuals or other stakeholders in joint fact-finding (Susskind and Field, 1996; Susskind et al., 1999) to collect all relevant information. By clearly understanding the limits of what is known, it becomes possible for people to make decisions that account for uncertainty but are not extreme in one direction or another.

1.5.11 Don't Be Afraid of Conflict—Work Patiently Through It

Biosolids recycling is likely to continue to be a controversial subject, at least in some areas. As the survey showed, public uneasiness about biosolids is common and if some pressure is added (e.g., a deadline for a permitting decision), this uncertainty can develop into concerned opposition and open conflict. It is at these early stages that the biosolids manager's actions have the most potential impact. If there are no looming deadlines and there are clear opportunities for discussion, the level of conflict may be reduced by communications, dialogue, and careful listening by the biosolids manager. All too often, as citizens raise concerns and ask questions, biosolids managers become defensive. This heightens distrust and concern on the part of the public and conflict escalates. Biosolids managers who are not trained in conflict resolution may give up and abandon the project. The result is not helpful—no one has learned about sewage sludge management and biosolids recycling and everyone has been left with uncertainty or unresolved hurt. The conflict remains to taint future discourse on biosolids.

A better approach is for biosolids managers to learn more about conflict and be ready for it. Conflict is a natural part of human interactions and can lead to improved understanding and positive change. If and when serious conflict does arise, Sandman (2000) recommends that biosolids managers react to public outrage as follows:

- "stake out the middle, not the extreme;
- acknowledge prior misbehavior;
- acknowledge current problems;
- discuss achievements with humility;
- share control and be accountable;
- pay attention to unvoiced concerns and underlying motives;
- leave views at the door; go into it to problem-solve."

Susskind and Field (1996) encourage the use of what they call the "mutual gains approach," which is based on the following carefully worded principles:

- "Acknowledge the concerns of the other side;
- "Encourage joint fact finding;
- "Offer contingent commitments to minimize impacts if they do occur, and promise to compensate knowable but unintended impacts;
- "Accept responsibility, admit mistakes, and share power;
- "Act in a trustworthy fashion at all times; and
- "Focus on building long-term relationships."

Other dispute resolution experts put it this way (Agency for Toxic Substances and Disease Registry, 1990): Accept and involve the public as a partner; plan carefully and evaluate your efforts; listen to the public's specific concerns; be honest, frank, and open; work with other credible sources; meet the needs of the media; and speak clearly and with compassion.

There is another factor to recognize about conflict. Sometimes, the biosolids manager will not convince the public. As conflict specialist Dr. Juliana Birkhoff (personal communications) notes: "You can have the best communications, excellent technical information…and the public could still say "no" and that's okay. This is a democracy." Biosolids managers should expect that this will occasionally be the outcome. While it may seem like a waste of time and money to go through a process of education and dialogue and still end up without a biosolids recycling project, the resulting relationships and trust developed in the process can be valuable in the future. Entrenched conflict happens. Sometimes, there is just no resolution, at least over the short term.

1.5.12 Monitor Public Perceptions and Public Relationships

*** a *most critical* action step**

As with private relationships, public relationships require routine maintenance. Once you have established decent relationships with stakeholders, you want to protect that investment of time and energy by keeping in touch on a regular basis. Maintenance of public outreach and participation does not take as much time and energy as establishing them—but it is important not to let it fall between the cracks.

Think of it this way: wouldn't you rather be communicating with stakeholders, the media, and the public when things are going well and no one is complaining? Wouldn't you prefer to be openly sharing information about the benefits and quality of your biosolids program on a routine basis rather than communicating only when something goes wrong or someone is upset and the focus is on the negative? If stakeholders, the media, and the public are kept up to date on what is happening with your program, they are more likely to understand the context of problems that occasionally arise and will be more likely to help find solutions and/or give you the benefit of the doubt.

Monitoring relationships means having some mechanism (e.g., a hotline) for hearing complaints or concerns as soon as they arise. You will be able to respond appropriately and in a timely fashion, showing that you understand the concern and are working to do something about it. Many people working at wastewater facilities and biosolids programs describe the benefits of being responsive and caring about neighbors' concerns—even if they can't do much in the short term to change the situation (e.g., make malodor stop). Just by expressing concern and understanding and explaining what might be done and what is being tried usually makes the concerned neighbor feel better. Of course it is critical to be accurate about what can and cannot be done, to set reasonable expectations, and to follow through on what you say you are going to do.

1.6 Building Trust

Biosolids recycling is currently plagued by deep distrust on the part of some vocal opposition and concerned citizens. Biosolids managers, public wastewater utility operators, biosolids research scientists, and biosolids regulators are all finding, sometimes to their surprise, that citizens concerned about biosolids recycling don't trust them.

In addition to the "outrage factors" discussed above (Table 1-3), other factors contributing to distrust of biosolids recycling programs include:

- the self-implementation aspect of the federal (and many state) regulations ("the fox is guarding the hen-house");
- the risk assessment basis of the regulations and management practices (risk assessment involves decisions and assumptions that are hard to understand and are subject to challenge because of potential bias);
- biosolids often come from somewhere else, from a much larger municipality whose people are seen as different and less trustworthy;
- biosolids are often conveyed by private companies that are perceived as having profit motives, making them less trustworthy;
- Class B biosolids are often given away, reducing their apparent value and suggesting there may be something wrong with them;
- rural communities often have little or no say in biosolids recycling decisions made in larger municipalities far away; and
- citizens sometimes receive little or no immediate response from a biosolids manager or farmer when they raise a concern or have a question about biosolids use.

As with other "outrage factors", biosolids managers can reduce the development of distrust by addressing these issues.

Above and beyond factors particular to biosolids management, there are personal attributes and actions on the part of the biosolids manager that will increase trust:

- honesty,
- openness (transparency),
- humility (not arrogant),
- compassion and caring,
- respectfulness,
- association with known, trustworthy people,
- say what you mean and mean what you say, and
- follow-through on promises.

These, and other trust-building attributes and actions, are the same as ones you use in all social interactions to gain the trust of others.

Finally, earning trust is different from gaining "acceptance" or "support." A stakeholder can have trust in a biosolids program while still not accepting or supporting it. He or she can feel confident that the program will operate in a particular way, that staff and management will do what they say they are going to do, etc.—and still not "support" the program. In some situations, the best the biosolids manager can do is to build trust, even if support remains elusive. Conversely, a lack of trust does not lessen the value of, and need for, good communications and predictable, respectful, trustworthy actions. Building trust takes time; losing it takes just one misstep.

Following are additional discussions of aspects of biosolids recycling programs and public participation that will impact the level of trust.

1.6.1 Remember, Most People Don't Know About Biosolids...

As noted above, most people don't know much about biosolids. Therefore, an educational process will almost always be necessary. These communications should begin early in a project. The alternative is to wait for the public to learn about biosolids on their own and risk the likelihood that they will form opinions around negative experiences; e.g., bad odors or a media story that stresses only the negative aspects of biosolids recycling.

How do you know where to start in sharing information about biosolids recycling? As this project's survey also indicated, public support for wastewater treatment is high—almost everyone supports the idea of cleaning up water that society has polluted. Thus, biosolids managers may increase the likelihood of a constructive dialogue about biosolids recycling if they—or, better yet, one of the more trusted information sources (government officials, academics)—reach out to the affected public early in the process to provide information about wastewater treatment and the sewage sludge that is a necessary result of that treatment.

1.6.2 ...And Are Uncertain When They Find Out

Some biosolids managers question whether or not the survey conducted for this project (see Chapter 4.0) really measured public uncertainty about biosolids recycling. They say that, in their experience, malodors are what trigger interest in biosolids, negative perception, and opposition. They are right that malodors often trigger initial negative reactions, as the literature and case studies clearly indicate. However, there are other odorous situations, such as manure spreading operations, that do not end up with the same level of public outrage, conflict, or concern. Why is this?

The survey responses clearly indicate an underlying uneasiness about biosolids recycling that makes the public response to malodorous biosolids different from the public response to malodorous manures. One clear reason for this is that biosolids are derived from human wastes, to which humans have an innate aversion developed as a requirement for survival. In addition, biosolids contain other trace contaminants that enter sewer systems. Biosolids are unfamiliar to people, and many will consider them "unknowable" because it is impossible to test for everything that might be in them. Animal manures are also unknowable, in the same way, but, overall, they contain fewer potential contaminants (manures don't have industrial influents), are not of human origin, and are far more familiar.

Public uncertainty about biosolids recycling creates additional concerns that the biosolids manager must be ready to address. Because of the complexity of biosolids' inputs, treatments, and management, there are hundreds of potential concerns about biosolids' impacts on public health and the environment. In some situations where staunch opposition has formed against biosolids recycling, these numerous potential concerns are raised one after another. This changing list of concerns presents a moving target for the biosolids manager, and the need to address concern after concern can be frustrating.

Even so, it is possible to pay attention to the larger audience's (the "guardians'") concerns and feelings, and address first those issues that are most critical to them. The biosolids manager will find that blatant—and even not so blatant—attempts by politicians or others to use an issue such as biosolids for self-gain are recognized by many people. If the biosolids manager continues to calmly respond to legitimate, important questions and concerns, most people are likely to develop respect and listen to him or her. Extreme views will become marginalized over time—but only if the biosolids manager works patiently to address concerns of most of the diversity of stakeholders. Remember that there is a continuum of diverse perspectives about biosolids. Assume each and every questioner is sincere in wanting to learn more from you; never discount or scoff at questions or concerns—most questions are reasonable to those raising them.

1.6.3 Integrate Local and/or Non-Technical Knowledge and Differing Perspectives

Biosolids managers are advised to value local and non-technical knowledge about the land and local conditions that may impact the biosolids recycling program. The fact is, people in local communities can provide a great deal of insight, critical perspective, and valuable details that can make a biosolids recycling program run more smoothly and avoid careless mistakes. Too often, biosolids programs have gone into communities without any communication with local people, only to be informed in an angry way at a public hearing about a new well or a seasonal wet area on the land application site—or some other glitch. Being informed about legitimate concerns in a public meeting quickly leads to lack of trust or respect for your efforts. As biosolids managers develop plans for new land application sites or scenarios, they can ask questions of the farmer, local political leaders, neighbors, and others and get a sense of local conditions that may be significant.

Most engineers, scientists, and regulators working in the biosolids management field are rational thinkers. When they enter into public forums, they encounter what they see as "irrational" processes. When irrationality is encountered, it is helpful to identify it as soon as possible, understanding that it may require a different approach. For example, if an extreme or irrational viewpoint is hampering constructive dialogue, it may be necessary for the group to isolate (marginalize) the individuals, negotiate with them to gain their constructive cooperation, or publicly expose the weaknesses of their arguments. While it is critical to never be disrespectful, it makes sense for the biosolids manager and other involved stakeholders to insist that public policy decisions, such as about biosolids recycling, be made using rational thought processes.[5] For the biosolids manager, however, the greatest danger in respectfully marginalizing a position is the real possibility that he or she will mislabel something or someone as irrational when in fact they are not. One way to avoid this is to consult with several diverse stakeholders involved in the process to determine whether or not your assessment of the extreme position is shared by others and that others agree that the constructive process is being hampered by that extreme position.

1.6.4 Identify Value Differences

Values underlie some of the conflicts over biosolids recycling. The most notable value differences in the biosolids/sludge conflict in North America are between those who focus on environmental issues, value "natural" things, support the Precautionary Principle, and hope for no degradation of the environment versus those who work to develop practical solutions to complex problems, value risk assessment, and feel that the benefits of modern society are going to cause negative impacts, although these should be minimized as much as reasonably (economically) possible (i.e., they value cost-benefit analysis). This value difference was evident in the conflict over the National Organic rule developed in the late 1990s by the U.S. Department of Agriculture (USDA). Because of biosolids' artificial, human, and industrial sources, more than 200,000 comments were submitted to USDA opposing the use of biosolids in organic production. Likewise, in Kern County, differences in values amongst the various stakeholders helped create the conflict (Chapter 3.2.13).

Elaine Vaughan, Ph.D., (personal communication) a specialist in public conflicts, suggests that biosolids managers entering into dialogue with communities should attempt to identify common values from which they can build their discussion of biosolids recycling. A dialogue that is grounded in identified shared values can lead to constructive discussions of risk perception. Finding areas of shared values prepares the public for closer examination of comparative risks.

At the least, biosolids public outreach and participation efforts should take into account core differences in values. The biosolids manager will gain credibility with environmentalists and communities if his or her actions are determined, in part, by consideration of what values are most important to stakeholders.

Core value differences cannot easily be changed. In most cases, the biosolids manager can only identify and accept them and understand that value differences do not preclude cooperation. Even knowing deep differences exist, it is still possible to work out solutions that everyone can accept.

---

[5] Note, however, that "rational" thinking does not include only science. As discussed below, there are other aspects of rational thinking, such as values. All are usually involved in public decisions.

1.6.5 Respect

"We always need to remember that the questions people ask which arise from genuine concern are valid and demand clearly understandable, open and honest responses," wrote Jane Forste (1995), a long-time biosolids manager and developer of public outreach programs. She was talking about respect. Biosolids managers and staff who respect differing points of view and recognize that there are no "stupid questions" will gain greater trust and respect in return.

It is hard to overstate the importance of acting with respect for others. In the U.S., the strongest, most vocal opponents were likely incited to hold their strong positions and remain motivated in their opposition in part because of their experience of not being respected, of being dismissed. Biosolids managers and the organizations who represent them must find ways to disagree with opponents' arguments without being disagreeable and without showing disrespect for the individuals themselves.

1.6.6 Size Projects Appropriately

The public responds with greater concern to projects that are perceived as large and unending. Planners of biosolids recycling programs can reduce public uneasiness by sizing projects appropriately. Consider and respect what the local community's perceptions and reactions might be. How great an increase in truck traffic, or dust, or farm odors would you be willing to have near your home?

With regard to a Class B land application site, for example, it may be better to submit a permit application for a smaller acreage—one field or two—than to include a large number of acres and fields all at once. This may increase the ultimate administrative effort (e.g., if additional permits are required for future expansion of the project), but it may make the difference between being able to go ahead with something and not being able to do anything.

Similarly, biosolids managers should determine how many years a site might be used and clearly state when the biosolids recycling program is expected to end (i.e. after how many years) and the likely schedule for each individual annual application event. Another option is to work with local citizens to gain support for an initial, smaller-scale, trial period, after which a reassessment of the project is conducted before it continues and/or expands. Many state environmental agencies are willing to provide permits for field "trials," recognizing that trials can help develop public experience and confidence.

1.6.7 Siting

Proper siting of biosolids processing and land application programs also affects the level of trust and public support. Sometimes biosolids recycling programs will settle for a marginal site in order to keep the biosolids moving. Any short-cuts of this sort are damaging to public trust (it's probably better, in the long run, to bear the extra cost and go to a landfill).

The siting of facilities that may be perceived as dangerous or, at the least, creating nuisances, has been a well-studied topic of social science research. Whitcomb et al. (1994) reported on the results of a public attitude survey conducted in the fall of 1993 to determine attitudes towards siting a hypothetical composting facility, noting the following:

- Greater distance from the proposed facility, a perceived increase in economic opportunity, a "waste involvement measure," and (male) gender positively influences the acceptance of a MSW facility in the respondents' community.
- On the other hand, perceived environmental impacts, lack of trust, and higher income of respondent increase the likelihood of a negative opinion of the proposed facility.

- Important conditions that respondents would like to require of the proposed facility are:
  - Facility monitoring and control procedures;
  - Written specifications as to the future responsibility of the facility; and
  - The community having the power to shut down the facility if problems occur.
- Important Host Community Benefits that were chosen by respondents include:
  - Guaranteed water quality,
  - Property tax credits, and
  - No charge for waste disposal (a clear local benefit).

Lessons from this and similar studies can be applied by biosolids managers to biosolids processing facilities and land application sites.

Susskind (1984) similarly notes that the most impacted individuals—and those taking the greatest perceived risk in accepting some new facility or land application site in their community—are those that are closest to the site. Thus, he suggests, it is reasonable to agree to mitigate or compensate these closest neighbors, if any significant negative impacts do occur. This can be done with "Host Community Benefits," as noted by Whitcomb et al. (1994), or may be even more closely targeted to the closest neighbors. While biosolids managers have tended to be unwilling to provide compensation to neighbors, pay host fees, or post performance bonds, these are all commonly used tools for addressing the different levels of risk and potential impacts experienced by close neighbors versus more distant community members. In short, if you can work out a fair way to make the close neighbors happy, then the strength of the opposition is going to be significantly reduced. People tend to fight hardest when something is close to them and they perceive they may have the most to lose.

1.6.8 Oversight and Enforcement

Over the past several years, a number of significant reviews of the federal U.S. EPA biosolids management program have identified lack of oversight and enforcement as a major problem. U.S. EPA systematically "disinvested" from its biosolids program after the promulgation of the Part 503 regulations in 1993. Given other priorities, the agency curtailed its program to the extent that the U.S. EPA Office of Inspector General (2000, 2002) identified oversight and enforcement as a major problem and major public concern. This concern has been echoed across the nation at public meetings concerned with biosolids recycling. Although the Clean Water Act provides for stiff penalties, the public is not comfortable with the honor system when it comes to biosolids management. And although many states do a great deal more oversight and enforcement than does U.S. EPA, it is still difficult for state and federal agencies to accurately answer questions such as:

- Where exactly were Class B biosolids land applied in the state last year?
- How about the year before?
- How many tons of biosolids were land applied in your jurisdiction?
- What were the levels of arsenic or some other pollutant in these land applied biosolids?
- Are you sure the required setbacks were followed?

Why is oversight and enforcement so important in the mind of the public? Because biosolids recycling involves a large number of potential concerns, including (but not limited to) "heavy metals," chemicals, pathogens, and nutrients. Citizens want to know that someone knowledgeable is making sure that potential risks are being managed properly. They know, as all biosolids managers should know, that the safety of biosolids recycling programs is dependent on proper treatment and management of biosolids from pretreatment through land application. U.S. EPA and other governmental agencies state that biosolids recycling is relatively safe—if applicable laws and regulations are followed. The public is very interested to make sure that "if" happens.

It is time for the biosolids recycling industry—from industrial pretreatment programs to wastewater utilities to biosolids managers to end-users of biosolids—to thoroughly and openly demonstrate compliance with all laws and regulations and attention to all best management practices. This is one of the basic premises of the National Biosolids Partnership's Environmental Management System program (see more on that, below).

There are two ways of going about this.

The first, more traditional way is to ensure that the state(s) in which you operate has (have) enough staff and funding to allow independent state inspections of your operations. In Washington State, the biosolids industry and public wastewater utilities worked with the state Department of Ecology (DOE) to develop a fee on each ton of recycled biosolids; the funds raised support DOE's robust monitoring and enforcement program. Another option is for public wastewater utilities and biosolids managers to lobby hard for increased state funding for the state biosolids oversight and enforcement program. The same could be done at the national level, although it seems unlikely that U.S. EPA will substantially increase its biosolids program budget, given other priorities and the fact that states are now looked to as the most reasonable and likely enforcers.

The second way of developing robust, open oversight and enforcement is for biosolids recycling programs to invite public stakeholders to assist. As you develop a new program in a new community, negotiate a systematic program with local officials and concerned citizens that involves them in oversight and enforcement. Work out agreements on operating and performance standards (e.g., odor limits). Invite citizens to communicate concerns through a complaint line or through regular phone calls or meetings. Another option is to provide a user fee that will pay, or partially pay, for a capable independent third-party person or town official to monitor how you are doing; this kind of approach has helped the Washington, D.C. biosolids recycling program for many years. There are program quality certification programs in many fields, and there are independent for-profit and non-profit organizations that provide quality assessments. The National Biosolids Partnership has quality certification in its Environmental Management System program.

Once you know your program is well monitored, be sure to let people see the results of the monitoring. Set up a website to post monitoring reports, hold meetings, provide details in annual reports, get it in the newspapers, etc. Don't be afraid to openly share failures along with successes—the public does not expect perfection; they do expect honest efforts and an overall record of quality. If you try to "spin" the monitoring story, you will lose credibility and trust. Then excellent reports will be distrusted. Just decide on a protocol for reporting monitoring information and stick to it.

State and federal biosolids regulators have learned in recent years that promotion of biosolids recycling is not an appropriate role for those involved in compliance, oversight, and enforcement. However you publicize the results of oversight and enforcement of your program, be sure to keep it clearly separate from any promotion of your program.

Some biosolids managers complain about lack of oversight and enforcement on the part of U.S. EPA and/or their state. A more productive approach is to develop an independent oversight and enforcement program of your own design as soon as possible. An easy way to do it is by developing an Environmental Management System program through the National Biosolids Partnership (see below). Yes, it will cost something, but if you are creative and work with the community, local oversight and enforcement does not have to break the budget. And the return for this effort can be as significant as any single action you can take. It may make the difference between having a sustainable, stable recycling program and one that is subject to repeated conflict.

1.6.9 Study Stakeholder Concerns

At times, biosolids managers and the biosolids industry as a whole have been slow to understand what critics and concerned citizens have been saying. Yes, many of the criticisms of biosolids recycling seem off-base or even ridiculous, but some of the critiques are valid and provide useful information. Biosolids managers should ask themselves, "Is there something important in that critique? If so, is there something our program could do to make sure this is not and never is a significant issue for us?"

These are the issues that critics talk about today:

- trace metals and chemicals ("pollutants"—all that can conceivably come down the sewer pipe),
- pathogens (organisms that cause human disease),
- odors and other air quality concerns,
- oversight and enforcement,
- surface water and groundwater quality,
- soil and food/crop quality,
- transportation and trucking,
- economic viability,
- changes in demographics and changing expectations,
- emerging issues (e.g., the fate of new pathogens, pharmaceuticals), and
- uncertainty about what is in a particular biosolids product.

The rational approach to designing a beneficial and publicly supported biosolids recycling program is to reduce as many of these public concerns as possible. Of course, the biosolids manager must balance the need for public support against other significant aspects of the biosolids management program, such as technical feasibility and cost.

1.6.9.1 They will Come After Class A Too…

In recent years, biosolids management decision-makers have been noting public resistance to Class B land application programs and have made choices to switch to Class A programs. The rationale is that, in almost all states, Class A biosolids have few, if any, restrictions on their marketing, transport, or use. Generally, no site-specific permits are required.

While the regulatory structure makes sense from a technical standpoint, biosolids managers should not be lulled into thinking that concerned citizens and environmental groups will always support Class A biosolids. Class A biosolids potentially have all of the same issues as Class B, except for pathogens (and there are even concerns expressed about regrowth or recolonization of pathogens and the adequacy of Class A certification tests).

There does seem to be a real advantage, in terms of public acceptability and marketability, to Class A materials that are thoroughly processed and stable. For example, many mature screened compost and pelletized heat dried fertilizer products have strong markets and significant customer and public support. However, if a Class A product has malodor and an unappealing texture or appearance, it may illicit negative responses. And if a Class A product is used in large volume in a neighborhood in which no one has been informed (and especially if it has some odor), it may trigger concerns, just like a Class B product.

Even managers of Class A products may need to pay close attention to biosolids quality concerns, reducing as much as possible any of the objectionable aspects of the material and how it is transported, stored, and utilized. Even for Class A biosolids products, it is critical to establish constructive relationships with key stakeholders—especially customers—and listen to, understand, and respond to customer and community feedback.

1.6.10 Engaging Stakeholders is the Greatest Challenge

Biosolids managers and organizations have experienced some difficulty, over the years, in getting concerned citizens or environmental groups engaged in problem-solving processes around biosolids. Engaging the general public—people who are less directly interested in biosolids—has been even harder.

There are reasonable explanations why people are not willing to commit much volunteer time to working on biosolids issues:

- biosolids are not as clear and attractive an environmental issue as, for example, land preservation or wildlife protection;
- biosolids are a relatively small proportion of agricultural fertilizers and soil amendments and, therefore, may have lesser overall impacts than, for example, manure management; and
- the public has not been widely invited to participate in biosolids management programs.

Because environmental groups and local community members are the ones most likely to become engaged in biosolids issues when they arise, it is critical to find ways to engage them as soon as possible. Biosolids managers, state and regional wastewater and biosolids groups, and national organizations such as the Water Environment Federation and the Association of Metropolitan Sewerage Agencies might increase their outreach efforts to keep up communications and dialogue with environmental groups, as well as other stakeholders such as conservation and farming interests. Establishing relationships with these organizations and key individuals will make it possible to gain their assistance and immediate input when issues arise.

The biosolids management industry needs to become open to greater public input, actively pursuing it whenever there is an opportunity. Peter Sandman (2000) notes that "people are much less desperate to have input when it is clear that they can whenever they want. One of the signs of a good public involvement strategy is relative lack of interest in getting involved." This kind of "no news is good news" is not the same as having a program that no one knows about (i.e., that is "under the radar"). The only way a biosolids manager can know whether or not people want to be involved is by constantly checking with them and offering the opportunity.

One way of getting people engaged is by developing a time-limited project that either puts biosolids to use to tackle an important local problem (e.g., see Chapter 2.2.4) or is seen as an effort toward a significant shift in policy or in the management of biosolids. Other options are to involve the public in odor assessment work or in developing local research or demonstration projects—efforts that promise to help improve plant operations and/or reduce potential or real impacts on neighbors.

Examples of projects that have gained constructive public input focusing on biosolids *as a tool* to address an environmental need are:

- the collaboration between King County, Northwest Biosolids Management Association, and the Mountains-to-Sound Greenway that utilized biosolids to revegetate areas scarred by past logging activity and protect land (see Chapter 3.2.1);
- construction projects where biosolids compost is seen as a tool to meet the need for minimum soil organic matter content that ensures, infiltration, reduced runoff, and surface water protection; and
- the restoration of mine spoils and mining scars where acid-mine-drainage has created toxic conditions and biosolids serve to reduce the mobility of metals from soils and stimulates stabilizing vegetative growth (e.g., at Kellogg, ID; Leadville, CO; and in Pennsylvania).

Examples of projects that have gained considerable public input around a significant policy or biosolids management concern are:

- Brattleboro, VT's collaborative outreach and planning project involving businesses, citizens, community leaders, environmental groups, and wastewater utility staff in addressing influent quality by stressing the need for awareness about proper disposal of (mostly household) hazardous waste;
- the National Biosolids Partnership's work on developing a system of credible independent audits for its Environmental Management System (EMS) program; and
- the developing collaborative research project on air emissions from biosolids land application sites by the PA Department of Environmental Protection, U.S. EPA, and USDA.

The latter two projects have clearly shown that interested citizens are willing to volunteer extensive time and energy—but only if they are provided with opportunities to provide significant meaningful input and be involved in decisions. For example, public interest and involvement can be energized by opportunities to help wastewater treatment facilities (WWTFs) to manage regional water, nutrient, and organic matter resources. In some locations, WWTFs are becoming recognized as major players in watershed management because they control large amounts of water and often have funds to support watershed projects. Participation of WWTFs in watershed issues can help to ensure that stakeholders address biosolids management needs as well.

In summary, as the wastewater treatment operator and biosolids manager face issues, he or she might consider if the issue offers an opportunity to engage any or all stakeholders in helping address the issue. By framing the issue as a community concern and defining a clear process by which it will be addressed (including a time limit, so people know they are not committing to an endless amount of work), it becomes likely that public stakeholders will become involved in a process that builds dialogue, mutual respect, trust, and relationships.

1.6.11 Information Networks and Regional Biosolids Groups

In most instances, biosolids managers are busy people, and they may have trouble finding time to follow a lot of the recommendations provided here. One source of support is to participate in a regional biosolids partnership or information network. The oldest and most successful regional association is the Northwest Biosolids Management Association (NBMA). This organization has developed, over more than 15 years, a strong cooperative network that exchanges information quickly and efficiently, supports cooperative research and demonstration projects, and aids all biosolids managers in the region with timely information, technical support, and outreach to stakeholders and the public. Active regional associations have also developed in California, Colorado, the Great Lakes, Mid-Atlantic, and New England regions. These groups serve as information networks and encourage cooperative efforts to address common issues.

Because the success of biosolids management programs often depends on local interactions, it is beneficial to have people available quite close to the local level, people who know the region, who can answer questions for local officials and citizens. A New Hampshire town leader, for instance, will not be as trusting of a fact sheet from U.S. EPA or a call from the National Biosolids Partnership as they will be talking with someone who is knowledgeable about biosolids and knows their state. Another benefit of a staffed regional group is that there is always someone in the region to address concerns quickly and with sound information.

1.6.12 You're Not Giving Up Responsibility or Control

The new emphasis on public participation encouraged in this report does not mean turning over control to non-professionals. It is still the responsibility of each wastewater treatment facility and/or biosolids manager to balance the needs of the various stakeholders for whom he or she works, including users of the sewerage system and biosolids end-use communities (e.g., farmers and neighbors). It is still his or her responsibility to ensure that biosolids are managed in accordance with current scientifically-based practices, thus protecting public health and the environment. It is his or her responsibility to represent and strongly state these interests in public participation forums and debates.

Biosolids managers are (or should be) experts because of education and experience, and the public relies on you for your expertise and advice. Experiences have clearly shown that citizens who become involved in reviewing and giving input to biosolids programs respect and seek the advice of biosolids managers and other experts in the field (e.g., see Chapter 3.2.6).

Working with interested stakeholders will enhance your expertise and experience, but you will remain ultimately responsible for your biosolids management program. As you invite the public into your information-sharing and decision-making processes, be sure to openly discuss this fact. Tell participants at the beginning what they can expect as a result of their participation and input. Will they be making the final decisions? Maybe not. But they may have significant participation and input in decisions, and it should be clear how that input impacts final decisions. If members of the public enter into a public participation process with clear expectations of their roles and the limitations of those roles, they will not be surprised and upset—or become distrustful of you and the process—if all of their input and/or recommendations are not adopted.

1.7 Risk Management and Program Vulnerability

1.7.1 Viewing Biosolids Management from a Risk Management Perspective

Just as risk management is a critical part of decision-making in developing wastewater and biosolids treatment infrastructure, so it is a useful lens with which to look at public outreach and participation.

Biosolids management programs are vulnerable to many external factors: weather, influent quality, etc. But often overlooked factors are the public: the community in which a program operates, the neighbors, and local environmental interests—even though public concerns sometimes create the most costly and intrusive disruptions to biosolids management programs.

The literature, case studies, and survey results suggest that, by increasing public outreach and participation, biosolids managers are likely to reduce program vulnerabilities and the risk of sudden costly setbacks for their biosolids management programs. Why? Because the biosolids manager will have better understanding of the public perceptions of his or her program, some public support, and knowledgeable and constructive public input to address any issue that arises.

In the private sector, corporations pay close attention to assessments of vulnerability and risks associated with the products they market. This has led to large corporate expenditures to address such issues as:

- worker health and safety in foreign manufacturing plants,
- total environmental impacts of manufacturing processes, and
- meeting public demands for corporate responsibility.

From corporate shareholder meetings to environmental groups to business organizations, there has been considerable pressure on large corporations to act responsibly and reduce their vulnerability to legal troubles (suits, regulatory actions) and public perception concerns (e.g., being labeled a "corporate polluter," "corporate sweatshop," or being accused of "greenwash"). For example, clothing manufacturers such as Patagonia and The Gap have spent millions to establish strict expectations and contracted protocols for off-shore clothing manufacturing facilities, in order to ensure fair treatment of workers and minimal impacts on local communities and the environment.

Manufacturers in numerous industries, from forestry and lumber to dairy products, have established environmental management systems and independent certifications that encourage managers and employees to take environmental impacts into account in decisions. Besides the fact that these programs can result in cost savings through increased efficiencies, they also serve to reduce vulnerability by addressing consumer and public concerns proactively.

There is evidence that insurance companies and courts treat corporations more favorably if the corporation has taken real, measurable, and proactive steps toward reducing vulnerabilities and liabilities. Recent prominent examples in the U.S. are the nascent efforts by a major food manufacturer and several fast food restaurant chains to reduce the fat and cholesterol contents of their foods while increasing their nutritional value as ways of reducing the potential of being liable in future lawsuits by affected consumers. These are all ways of managing risk and vulnerability in a competitive world.

Reducing vulnerability also makes sense for public utilities. Biosolids managers are encouraged to keep a running, prioritized list of both technical issues and public concerns and perceptions of their biosolids management program. The technical aspects are important, but the focus on this report is on the potential vulnerabilities regarding public perceptions.

An initial list of potential vulnerabilities can be developed by considering items listed in Section 1.4 and in the twelve action steps in Section 1.5. The list of vulnerabilities then needs to be prioritized so that the biosolids manager can focus on the greatest vulnerabilities first. For example, as the survey and literature review indicate, one of the most widespread and significant concerns of the public is the organic and inorganic chemical quality of biosolids products. This vulnerability can be addressed proactively in many ways:

- increase the comprehensiveness of industrial pretreatment programs (be creative in working with industrial and commercial sewerage system users to constantly improve pretreatment);
- step up inspections and enforcement of industrial pretreatment programs,
- invest in and partner with service area hazardous waste management programs to reduce opportunities for improper and/or illegal disposal to sewers or septic systems;
- encourage replacement of consumer products that are known to cause significant negative impacts on biosolids quality (for example, in the 1990s, some utilities identified baby diaper-rash creams as a significant source of zinc in biosolids and worked with cloth diaper washing services to provide consumers zinc-free creams at no-cost to the consumer);
- increase the frequency of biosolids quality testing (many larger producers concerned with biosolids quality for consumer markets test more often and for more parameters than required by regulations);
- have an independent statistical analysis of biosolids quality data with regard to trace chemical and elemental contaminants (the State of New Hampshire, for instance, contracted with a university professor to run statistical analyses of biosolids data to allow statements such as "The chance that a New Hampshire Class B biosolids product will have a level of chromium that exceeds the regulatory limit is less than 5%");
- work with environmental and community groups to identify chemicals and elements of greatest local concern and work with those groups to address those concerns (for example, get involved in efforts to replace an older medical incineration facility with a more modern and environmentally-friendly medical waste disposal system in order to reduce airborne distribution in the service area of mercury, dioxins, and furans);
- publish test results and provide open, frequent communications with the public about biosolids quality and the steps taken to protect it.

The actions a biosolids manager chooses to take will depend on an estimate of the potential impact of each action on reducing actual and perceived vulnerability and how much each action will cost in money and staff time.

As is evident in the list above, a biosolids management facility's interests in addressing critical vulnerabilities such as biosolids quality may result in actions beyond the "normal" scope of operations. These activities, however, may be the most efficient way to reduce a vulnerability. As an example, some research indicates that trace elements, such as those widely regulated at the state and federal level, can be removed from biosolids at a treatment facility. But what is the cost of installing and operating such a system in comparison to taking actions upstream, in the community, to achieve the same level of trace element reduction? Cost-benefit analyses can be done to address such a question.

In the water side of wastewater treatment facilities, the use of chlorine has often been replaced with ultraviolet radiation for disinfection of effluent, in part due to concerns about the vulnerabilities associated with storing and using chlorine. Therefore, public utility biosolids managers might increase their knowledge of the polymers used for biosolids treatment, how they are produced, and what alternatives exist.

Another possible avenue for biosolids managers to reduce vulnerabilities is to evaluate the vendors they work with. Evaluating vendors' practices, reputations, and product qualities is becoming standard practice in corporations that have EMS programs. In many situations, before they contract with a vendor, private corporations will ask the vendor to complete and certify their answers to questionnaires about their product quality, business practices, treatment of employees, and environmental record. If the vendor marketplace is highly competitive, some corporations go so far as to require that the vendor they choose has a robust environmental management system and provides routine reports on the development, goals, and accomplishments of that program. The goal is to reduce the potential of utilizing products that may vary in quality and/or to reduce the potential vulnerability of having the corporation's products associated, at some future date, with a vendor whose products or practices are the target of public outrage and legal challenges.

Perhaps most critical for public utilities' biosolids management programs is an evaluation of the contracted hauling and use or disposal of biosolids. Private biosolids management companies manage a large portion of the biosolids beneficially used on this continent. Most utility biosolids managers already pay close attention to biosolids management companies' activities, ways of doing business, quality of staff, and reputations.

But, in many cases, these assessments are not as thorough as they could be. Some contracts have only a few specified requirements regarding the quality of the biosolids management program, including such things as having back-up use or disposal options at all times, ensuring sufficient short-term storage capacity, addressing trucking issues and potential impacts on communities, and detailing best management practices for land application. Many contracts with biosolids management companies do not include anything about public outreach and public participation.

Through the NBP EMS for biosolids program (Section 1.7.2), attention is directed to the quality of the entire "biosolids value chain"—all those aspects and inputs to a biosolids product and program that affect biosolids and program quality, environmental impacts, and public perceptions and support. Some public utilities are beginning to rewrite their standard requests for proposals (RFPs), bid documents, and contracted services contracts to incorporate greater attention to overall product and program quality. They are dictating higher standards for private contractor operations.

Encouraging this approach throughout the biosolids management industry (e.g., by sharing these kinds of enhanced documents and encouraging their wider adoption) will help advance the overall quality of biosolids management programs by ensuring that those few private contractors that operate by cutting corners and ignoring public concerns are weeded out. As part of this approach, public utility biosolids managers need to ensure there is a robust system of accountability detailed in contracts and that the utility regularly checks on contractor performance (including, for example, how they approach local communities). Some public biosolids managers have considered requiring the biosolids management company with whom they contract to develop its own EMS.

1.7.2 Environmental Management Systems

An important means of reducing potential program vulnerabilities is to adopt an environmental management system (or other enhanced management system) that focuses on continual improvements in overall product and program quality. The National Biosolids Partnership's (NBP) Environmental Management System (EMS) is based on the Deming quality management model of constant improvement. An EMS program can be an organized, rational, step-by-step process that leads to the reduction of vulnerabilities and liabilities in all aspects of a system or program.

A good environmental management system program, such as that of the NBP, allows considerable flexibility and tailoring to the local situation. Thus, there are a limited number of requirements for what has to be in the NBP EMS. Instead, local biosolids managers determine the focus of their EMS and its actions through a step-by-step cyclical process of assessment, prioritization, action, and evaluation. Key components of the NPB EMS include (National Biosolids Partnership, 2002b):

- **Independent program audit**, including tracking of actual environmental impacts of the biosolids management program with the intent of reducing negative impacts.
- **Public participation (Element 6):** EMS guidance requires organizations to "select and implement a proactive public participation approach to involve interested parties in its Biosolids Management Program and EMS planning process." The approach must include a plan for third-party verification of conformance with NBP standards.
- **Communications (Element 9):** Organizations are required to "establish and maintain a proactive Communications Program that provides ongoing information about its Biosolids Management Program and its EMS to interested parties and the public, consistent with local circumstances, the method of biosolids management, its public communications history, and degree of current interest in its biosolids management activities." The program also must "make available a summary of its independent, third party EMS verification audit results to the public."

To become a certified NBP EMS program, the biosolids manager and staff are required to go through a rigorous assessment of vulnerabilities ("gap assessment"). They then prioritize those vulnerabilities and commit, on paper, to constant improvement through addressing each vulnerability, starting with those of highest priority. At least once a year, the program is evaluated, its successes and failures are noted, and new commitments for the next year are made. Once the EMS has been in operation for at least six months, an independent audit can be requested by the utility. If the audit results are satisfactory, the EMS becomes certified by the NBP.

The NBP EMS program requires a level of "transparency" and public participation that was developed with input from several environmental group representatives expert in environmental management systems. Thus, having an NBP-certified EMS for biosolids should enhance public participation and support and serve to reduce liability and vulnerability around biosolids management. Aspects of the program that build trust and enhance public perceptions are:

- The EMS requirement of continual improvement, which encourages *a willingness to make changes based on public input and participation*.
- The EMS emphasis on attending to details at all localities where the biosolids management program interacts with the public, which encourages *accountability at the local level*, where the biosolids meet the soil.
- The EMS emphasis on covering the entire biosolids value chain, which encourages biosolids managers to adopt a *systems thinking approach*. In relation to communications and public participation, a systems thinking approach can lead the biosolids manager to consider the impacts of each part of the biosolids value chain on public perceptions and relationships (it may be useful, for example, to diagram the biosolids value chain and identify and record the likely impacts of each part of the value chain on public perceptions and relationships).

Just as significant, a few new and developing NBP EMS programs have already experienced some cost savings through the identification by the EMS process of ways to increase efficiency. For example, the Washington, DC Washington Area Sewerage Authority estimates it will save a million dollars annually through an efficiency in biosolids treatment that, the operators believe, would not have been identified without the EMS process.

The NBP EMS program for biosolids is not the only way for public wastewater utilities and biosolids management organizations to develop quality management or an environmental management system. In fact, a few public wastewater utilities (e.g., Lowell, MA) have developed EMS programs under the internationally recognized ISO 14000 standard. Those facilities that have gained ISO 14000 EMS certification have achieved cost savings and reduced environmental impacts through such aspects of their operations as pump efficiency and saving on electrical costs, which also reduce the facility's need for energy and its related impacts on the environment. These kinds of savings and reduced environmental impacts are less likely to be realized in the NBP EMS program, because of its narrower focus. However, an NBP EMS is less cumbersome and costly to create and operate, an important consideration for public utilities whose budgets are tight. In addition, the detailed process of the NBP EMS can start the utility or company management and staff on the environmental management mode of thinking, which means that the same process can subsequently be applied to other aspects of the wastewater utility or company operations.

1.8 Addressing Conflict at the National Level

1.8.1 Public Participation in Development of Policy

In the U.S. and Canada, the development of public policy usually includes significant requirements for public review and comment. This is a natural development in any democratic society. However, public review and comment processes often are streamlined in an attempt to balance the needs of those proposing something and those who may wish to comment.

Biosolids managers are used to public hearings and meetings to inform the public of actions that may impact them and that offer them a chance for review and comment. Most biosolids managers also are well aware of the shortcomings of such meetings—they can increase the dissemination of erroneous information, increase extreme positions, and increase conflict. There are several reasons for this:

- Public hearings are often the one and only chance for the public to vocally comment on a proposal, so they feel pressured to strongly state their position;
- Public hearings are usually scheduled soon before a final decision is made, which leaves the public little time to learn about the proposal, ask questions at the hearing, and/or provide substantive comments;
- Public hearings are large group meetings where experts usually stand in front of "non-experts," creating an imbalance of power and authority to which the public can react with uneasiness and outrage;
- Public hearings tend to encourage simplified, vocal, exaggerated arguments and grandstanding more than conscientious, constructive dialogue that leads toward better mutual understanding; and
- Because of the above, public hearings are often dreaded by presenters and many others involved in them, which heightens tension.

Despite these shortcomings, social scientists who study constructive public participation in the development of public policy agree that the legally required public hearing and comment processes are necessary. What these researchers encourage, however, is the use of informal processes, such as consensus building, *in advance of* or *parallel to* the formal, legal process.

Biosolids managers are encouraged, for example, to hold smaller meetings with neighbors and others in the community and take other proactive relationship-building steps in advance of a required public hearing. In addition, people developing regulations are encouraged to consider alternative dispute resolution processes, such as consensus building, to address conflicts and come to adequate resolutions in advance of schedule-driven, formal policy-setting processes.

The time has come for the biosolids management industry to consider ways to address the growing conflict over biosolids policy at the national level. Experience over the past several years has clearly demonstrated that this conflict is having a significant impact on local biosolids management decisions in many parts of the country (e.g., see Chapter 3.2.13). Decisions being made in the midst of heated conflict may not be in the best interests of society or the environment. Creating a forum for dialogue amongst people representing the wide variety of viewpoints may be necessary to avoid the costly mistake of hasty decisions driven by public outrage or reactions to public outrage.

In order to create a meaningful public policy dialogue, it will be incumbent on leaders in the biosolids management field to find ways of involving a wide variety of interests, including concerned citizens (but not just the most concerned opponents). Environmental, agricultural, and public health organizations should be involved, as well as representatives of the stakeholder groups currently active in biosolids management (e.g., farmers, research scientists, regulatory agency staff, and biosolids managers). Getting this variety of groups involved is a significant task and will take considerable time, money, and effort. It is likely that such an effort will only succeed at a national level if the national leading wastewater and biosolids organizations commit to it.

Models for diverse public participation that have been developed around other controversial issues may be useful for biosolids management. For example, a national stakeholder group was formed in the mid-1990s to help the wind energy industry deal with public concerns about the siting of windmills and "wind farms." This group, the National Wind

Coordinating Committee (NWCC), has had as many as 50 members (some only work in one or more of five smaller workgroups). The NWCC facilitates decision-making around wind energy facility siting, technical issues, research agendas, and more. Stakeholders include wind energy development companies, financial institutions, landowners, neighbors to wind farms or proposed wind farms, and environmental groups. The NWCC is organized and facilitated on an ongoing basis by Resolve, a Washington, D.C.-based nonprofit that specializes in conflict resolution and consensus building. Resolve has two full-time staff, one half-time secretary, and a half-time mediator supporting the NWCC. The NWCC runs workshops around the U.S. and jointly develops informational products. Much of its funding is from the U.S. Department of Energy, which provides about $500,000 a year. Evaluations of the NWCC's work have shown considerable support and appreciation from diverse participants (National Wind Coordinating Committee, 2002 and personal communication).

What seems clear, is that, in the absence of any further concerted efforts to address the national conflict around biosolids recycling (especially around Class B land application), the national conflict will continue to negatively impact local biosolids management programs. Which is more costly, the accumulating local restrictions on biosolids recycling programs and associated costs of litigation or a proactive concerted effort to address the conflict and public concerns on a national (or possibly regional) scale? It will take some expert analysis to find the answers to these questions.

And while there have been some experiences with dialogue amongst parties in the biosolids conflict (e.g., the so-called "New York Roundtable") that have been seen as failures, there are many examples of situations in which biosolids managers and concerned members of the public have found things they can agree on and work on together (e.g., see the first few case studies in Chapter 3.0 on the accompanying CD-ROM). The same could be done on the national or regional level.

As Lawrence Susskind emphasizes, it is best if those in the biosolids management field always strive to learn more about biosolids and about biosolids management. Part of this striving for more information and understanding is learning from those with different points of view.

1.8.2 Public Participation in Biosolids Research

The National Research Council (NRC) report of July 2002 recommended stakeholder involvement in biosolids research and risk assessments. Why? Because decisions about what to study and what assumptions to apply are necessarily biased by one's experiences and perspective. Therefore, when making such decisions it makes sense to ensure representation of the variety of perspectives and experiences related to biosolids management.

In response to the NRC report and other similar recommendations, U.S. EPA, USDA, and the Pennsylvania Department of Environmental Protection (PA DEP) created an Information Sharing Group (ISG) of concerned citizens to be involved in a large cooperative study of air emissions from biosolids and other land application sites. Development of a quality assurance project plan was undertaken in 2003. The ISG has provided useful input to the development of the research process protocols. However, there has been tension and some of the scientists have found that public input requires more work. For their part, the ISG participants are understandably frustrated in their limited role of providing comments and asking questions; they were not involved in the original design of the project, in funding, or in project staffing decisions. Thus, this formal involvement of concerned public citizens in a biosolids research project did not match the ideal model for public participation. But it was a significant step in the direction of increased public participation in research in the biosolids field.

Similarly, the Water Environment Research Foundation (WERF) has decided to bring public input into its research processes, especially in areas, such as biosolids, that are the subject of considerable public interest. WERF created a July 2003 national biosolids research summit to help determine a biosolids research agenda for the next five or more years—and, for the first time, this biosolids research agenda-setting process included diverse public input in the planning and production of the summit, as well as at the summit itself.

These developments are grounded in the principle of "joint fact-finding" (Susskind et al., 1999)—people with differing opinions and perspectives working together to establish mutually-agreed-upon facts and data. Over the past 10 years, this aspect of alternative conflict resolution has become a significant subject of social science investigation and experimentation. Joint fact-finding offers a means of avoiding "dueling science," where people on each side of an issue cite those researchers and scientific papers that they believe support their position. Although debate over data and conclusions is part of the scientific process, dueling science in a policy debate leads to increased public uncertainty about where "the truth" lies. It confuses policymakers and the public. It adds to divisiveness, reduces trust, and reduces the possibility of constructive, long-term relationships.

Instead of working on the science separately, "joint fact-finding" or "collaborative research" involves working on the science *together*:

- Work together to determine research needs, establish a research process that is credible and technically sound in the eyes of all involved (or almost all), and thereby establish credible data. In short, develop a common data and knowledge base.
- It is understood from the start that the way in which the data are interpreted and applied in policy may differ among the parties, depending on their viewpoint—and this is okay.

The benefits of joint fact finding include:

- Working together increases the scientific knowledge and understanding of all participants.
- Joint fact finding helps identify the most accurate objective information and its limitations.
- Working together means that the process of coming up with the data is known to all, witnessed by all, is much more "transparent," and thus less likely to be misunderstood or the target of criticism.
- Joint fact-finding can be helpful in increasing understanding of scientific uncertainties and how they may be dealt with.
- Working together provides the possibility of enhanced dialogue, mutual understanding, greater sense of fairness, increased trust, and improved relationships.

1.9 Conclusion

Biosolids management in North America may be at a critical juncture. Either biosolids recycling programs will develop better understanding, communications, dialogue, and relationships with the public—or the expansion in biosolids recycling of the past several decades may end amidst increased public outcry.

The authors of this report have learned from experiences in other fields that there are concrete actions that can be taken to

- improve public perceptions and acceptance of biosolids recycling and
- improve the quality of biosolids management programs.

Such actions are not new and include the same actions required in all relationships: showing respect, ensuring fairness, developing trust, establishing two-way communications, sharing appropriate information, and showing commitment.

Biosolids managers can work today, through small or large action steps, toward improving their public relationships with the diversity of stakeholders involved in or affected by their biosolids management programs. There are many tools and resources available—not just in this report, but in workshops, conferences, books, and other publications.

As you proceed , remember the one key question you can ask yourself in any situation before you take action or say something:

> *"What will be the impact of my next action on our program's relationship(s) with key stakeholders?"*

If the impact will be negative, think of something better to say or do.

REFERENCES

Adams, Paul L. 1998. The Blue Ribbon Sludge Program, A discussion of the status and organization of the sludge program in New Hampshire with recommendations for changes and improvements. Unpublished paper.

Adkins, Gary. 1975. 'Outstanding Project'–But Sludge Stinks, Neighbors Say. *Illinois Issues,* August. (Available at http://www.lib.niu.edu/ipo/ii7508234.html.)

Agency for Toxic Substances and Disease Registry. 1990(?). *A Primer on Health Risk Communication Principles and Practices.* Public Health Service, U.S. Department of Health and Human Services.

Alexander, R. 1993. Market Surveys and Market Development. In *National Extension Compost Utilization Conference*, Minneapolis, MN: Minnesota Extension Service/Cooperative Extension System. 41-43.

Alpert, Joel. 1999. A Comparison of Biosolids, Manures, and Chemical Fertilizers. E & A Environmental unpublished report.

American Broadcasting Company (ABC) News. 1995. *Prime Time Live.* February 8.

American Water Works Association Research Foundation. 2001. *Guidance to Utilities on Building Alliances With Watershed Stakeholders.*

Ammons, N. and M.R. Hill. 2000. Public Participation Necessary. *Water Environment and Technology*. 12(9):79-83.

Anderson, Connie. 1992. Smell That "Fresh Air". *The Similkameen Spotlight.* October 28.

Anderson, Jack. 1999. Govt. Fails to Regulate Sludge Dumping. Syndicated column, *Laconia Citizen,* NH, January 2.

Associated Press, The. 1999. Amelia Bans Human Waste Fertilizer. *Richmond Times-Dispatch.* April 9.

Associated Press, The. 1999. Compost Odor Makes Farmers Target of Lawsuit. *The Associated Press State and Local Wire*, State and Regional section. May 24.

Austin, TX, City of. 2000. Dillo Dirt: An Award-Winning Program. Product description. http://www.ci.austin.tx.us/water/dillo.htm.

Bakersfield Californian, The. 2000. County Fights State Over Waste Disposal Guidelines. October 5.

Barnett, Megan. 2002. Making a Stink: Neighbors Say Sewage Sludge Fertilizer Makes Them Ill. *U. S. News and World Report,* August 5, pp. 48-50.

Baron, J., J.C. Hershey, and H. Kunreuther. 2000. Determinants of Priority for Risk Reduction: The Role of Worry. *Risk Analysis*, 20(4), 413-27.

Bastian, R. 1993. Where Do We Stand on Regulations? *Sewage Sludge: Land Utilization and the Environment*, Bloomington, MN: Soil Science Department and USDA-ARS, University of Minnesota. August 11-13.

Bedell, C. 2000. County fights state over waste disposal guidelines. *The Bakersfield Californian.* October 5.

Beecher, Ned. 1996. Regulatory Shift Impacts Biosolids Recycling in New Hampshire. *BioCycle*, February.

Beecher, Ned. 1999(a). Education, Advocacy at Heart of New England Association. *BioCycle* 40(8):75-78. August.

Beecher, Ned. 1999(b). Cultivating New England Biosolids Recycling. *Journal of the New England Water Environment Association*, Vol 33, No. 2. 136-157.

Beecher, Ned and William E. Toffey. 1999. Since Politics is Local, We Must All Hang Together. *Proceedings of the Joint Residuals Conference.* Alexandria, VA: Water Environment Association.

Behnke, Gordon and Tom Amundson. 1993. Using Biosolids in Your Backyard: Turning "NIMBY's" into "YIMBY's". *Entering a New Era.* Alderbrook, WA. Northwest Biosolids Management Association Annual Biosolids Management Conference.

Belkin, Lisa. 2002. The Odds of That. *New York Times Magazine,* August 11. pp. 32-61.

Berger, Ida E. 1997. The Demographics of Recycling and the Structure of Environmental Behavior. *Environment and Behavior*, 29(4), 515-531.

Berry, Michelle and Clifford W. Scherer. (1990). Utilizing Surveys as Community Participation Mechanisms: Toward a Citizen Empowerment Strategy. Presented at the First United States Conference on Municipal Solid Waste Management; Solutions for the 90s sponsored by the U.S. Environmental Protection Agency, Washington, D.C., June 13-16.

Billings, Loren. 1997. Citizens Win; Coleman Agrees to Withdraw Sludge Application: Owner Will Use Topsoil to Reclaim Gravel Pit. *The Conway Daily Sun,* NH, 9 (70), April 14.

BioCycle. 1997. County Bans Biosolids For Land Application.

Bisogno, T. 1995. The Sludging of America, Sewage Waste Spread on Farms and Landfills is Causing Chronic Health Problems. *E / The Environmental Magazine,* May/June: 19-22. www.emagazine.com. May-June.

Bleifuss, Joel. 1995. Nightmare Soil. *In These Times,* October 16-29: 12-16. A rebuttal by Alan Rubin, U.S. EPA, is available.

Blethen Maine Newspapers. 2001. Lyman's Sludge Debate is Likely to be Repeated. January 8.

Bosomworth, C. Myrna. 1992. A Clean Valley, Please. *The Similkameen Spotlight.* October 28.

Boston University School of Public Health, Department of Environmental Health. 2001. Press Advisory. October 29.

Braile, R. 1998. Environmentalists Blast State Plans on Waste Control. *Boston Globe, NH Weekly.* November 15.

Braman, Jonn. 1998. Princeton Will Be Main Show at Fall Symposium. *The Similkameen Spotlight.* April 21.

British Columbia Lung Association. 2000. Changes in Princeton Bring Breath of Fresh Air. May.

Brobst, R.B. 1994. Biosolids Project Acceptance; New York City's Biosolids are Land Applied to the Eastern Plains of Colorado. In *The Management of Water and Wastewater Solids for the 21st Century: A Global Perspective*, Washington, D.C.: Water Environment Federation. 9-13–9-24.

Brooks, John P., Charles P. Gerba, and Ian L. Pepper, 2004. Biological Aerosol Emission, Fate, and Transport form Municipal and Animal Wastes. Journal of Residuals Science & Technology: vol. 1, no. 1, January.

Buckley, S.W. and T.J. Muirhead. 1994. Composting Success: An Odor-free Beneficial Use Product. In *Proceedings of the Water Environment Federation 67th Annual Conference & Exposition, WEFTEC '94*, Chicago, IL: Water Environment Federation. October 15-19. 677-686.

Bunn, Michael. 1995. Reader wants Nutrifor Used On Blockades. *The Similkameen Spotlight.* September 13.

Burbridge, Don. 1998. Nutrifor takes naturalists on demo tour. *The Similkameen Spotlight.* June 10.

Burke, Thomas A. 2002. The Science of Recycling Sewage Sludge. Washington, DC: National Academy of Sciences. September 6.

Burros, Marian. 2000. U.S. Imposes Standards for Organic-Food Labeling, *New York Times*, December 21, 2000.

Bynum, James. 1999. EPA and the New Plagues. Copake, NY: National Sludge Alliance. March 31. http://www.purefood.org/toxic/toxicpoison.cfm.

Cable News Network (CNN). 1997. Hazardous Harvest: More Than a Bad Smell?, Denver Sludge is Stinky Biz, and EPA Defends Sludge Rules. *Moneyline,* special reports (television) by Bill Dorman, June 25, 26, and 27. http://cnnfn.cnn.com/1997/06/25/busunu/hazard_one_pkg/

California Water Environment Association. 1998. *CWEA Manual of Good Practice, Agricultural Land Application of Biosolids*. Prepared by the CWEA Steering Committee on Manual of Good Practice for Agricultural Land Application of Biosolids, Oakland, CA.

Calverton Citizens' Association, 1996. *Calverton Current*. Calverton, MD: Calverton Citizens' Association. December.

Calverton Citizens' Association, 1997. *Calverton Current*. Calverton, MD: Calverton Citizens' Association. February.

Canody, Jeremy. 1997. Improving the Public's Perception of Biosolids. *Small Flows*, 11 (4): 5-6. National Small Flows Clearinghouse.

Carlile, B.L. and T. Gillane. 2001. Applying New York City Biosolids on a West Texas Ranch – Eight Years of Operation and Results. WEF Joint Residuals and Biosolids Management Conference, San Diego, CA.

Center for Food Safety, 2003. Petition Seeking an Emergency Moratorium on the Land Application of Sewage Sludge."

CH2M Hill. 1982(?). Sludge Management: Public Perceptions and Politics. A CH2M Hill study conducted by Consumer Concepts, Milwaukee, WI.

Chaney, R.L. 1990(a). Public Health and Sludge Utilization. *BioCycle,* 31 (10): 68-73.

Chaney, R.L. 1990(b). Twenty Years of Land Application Research. *BioCycle,* 31 (9): 54-59.

Chaney, R.L., James A. Ryan, and Sally L. Brown. 1999. Environmentally Acceptable Endpoints for Soil Metals. In W.C. Anderson, R.C. Loehr, and D. Reible (eds): *Environmental Availability of Chlorinated Organics, Explosives, and Metals in Soils,* Am. Acad. Environ. Eng., Annapolis, MD.

Cherry, S.R. 2000(a). EPA's Secret Role in Toxic Sludge. *Insight on the News* 16(27):17. July 24.

Cherry, S.R. 2000(b). Sludge Excuses Still Stink. *Insight on the News*. May 29.

Cherry, S.R. 2000(c). Toxic Waste Used as Fertilizer? *Insight on the News*. May 15.

Chesapeake Occupational Health Services. 1991. Health surveillance program for compost workers: An epidemiological review. WSSC Site II, Silver Spring, MD.

Chess, Caron and Kristen Purcell, August 1999, "Public Participation and the Environment: Do We Know What Works?" *Environmental Science & Technology*, 33(16):2685-2692.

Chrostowski, Paul C., Sarah Foster, and Damian Preziosi. 2002. *Scientific Peer Review of "The Case for Caution."* Presented at the New York Water Environment Association. Takoma Park, MD: CPF Associates. February 3-6.

Chrostowski, Paul C. and Robert G. O'Dette. 2002. Demystifying the Great Biosolids Debate: Sound Science Removes Emotion From Decisions About Biosolids Recycling. *Pollution Engineering*, 34: 8. Business News Publishing. August.

Clark, D. 1994. "Boulder Shouldn't Dump Sludge in Gunbarrel." *Boulder Daily Camera.*

Clean Water Report. 2002(a). Both Sides in Land Application Will Take Views to the Public. Silver Spring, MD: Business Publisher's, Inc., 40(2):11-12. January 28.

Cleveland, Charles B. 1976. Chicago: Joanne Alter, MSD Commissioner, Turns Engineering Ideas into Everyday Words. *Illinois Issues,* January. Available at http://www.lib.niu.edu/ipo/ii760130.html.

Clines, Francis X. 2000. The 2000 Campaign: The Battle For West Virginia; King Coal Casts Shadow on Governor's Race. *New York Times*, October 27, 2000.

Cogger, Craig and Dan Sullivan. 1991. *Recycling Municipal Wastewater Sludge in Washington.* WSU College of Agriculture & Home Economics.

Cohen, Nevin, June 1995, "Technical Assistance for Citizen Participation: A Case Study of New York City's Environmental Planning Process," *American Review of Public Administration*, 25(2):119-135.

Connelly, Shelagh. 2000. Beneficial Use, Politics, and the Cost of Upgrades, a presentation to the New England Water Environment Association Residuals Management Conference, November 16.

Cooper, Chris. 1995. Save Our County from Perception. *The Biosolids Trilogy*. Silverdale, WA. Northwest Biosolids Management Association 7th Annual Biosolids Management Conference.

Couch, S.R. and S. Kroll-Smith. 1994. Environmental Controversies, Interactional Resources and Rural Communities: Siting vs. Exposure Disputes. *Rural Sociology*, 59(1), 25-44.

Covello, V. and P. Sandman. 2001. Risk communication: evolution and revolution, in A. Wolbarst (ed.): *Solutions to an Environment in Peril.* Baltimore, MD: John Hopkins University Press, 164-178.

Covello, Vincent and Steven Wolf. 2003. Three Key Equations for Solving Public Meeting Nightmares. *Water Environment & Technology*. Alexandria, VA: Water Environment Federation, February.

Culos, Ermos. 2000. Let's Not Take This Crap From GVRD. *The Ashcroft Journal*. August 8.

Currie, Beth. 2000. Open Letter to the Town of Princeton. *The Similkameen Spotlight.* July 4.

Dahl, Diana and Roy Smith. 1998. Agricultural Use of Sludge: "Don't" is the Only Safe Guideline. *Hawaii Environment & Health News,* Vol. 1, No. 1.

Daily Commercial News and Construction Report. 1994. Lining Up For Sludge. March 8.

Davis, Charles E. 1993. *The Politics of Hazardous Waste*. Prentrice Hall, Englewood Cliffs, New Jersey.

Denver Publishing Company. 1998. Contamination Plans, an Underreported Story. *The Denver Rocky Mountain News*. 41A. August 19.

Doherty, Lyonel. 1999(a). Biosolids Still Worry Residents. *The Keremeos Review.* January 21.

Donovan, John F. 1993. Getting the Sludge Rule On Line. *BioCycle,* April 1993.

Douglas, Mary. 1992. Risk and Blame (essay). *Risk and Blame: Essays in Cultural Theory*. Routledge, London and New York.

Dowd, Scot E., Charles P. Gerba, Ian L. Pepper, and Suresh D. Pillai. 2000. Bioaerosol Transport Modeling and Risk Assessment in Relation to Biosolids Placement. *Journal of Environmental Quality,* 29:343-348.

Downing, B. 1997. From Septic Tanks To Ohio Farmland. *Akron Beacon Journal*. http://Wayne-Health.org/septage_1297.html. December 14.

Draman, Grace A. 1994. The Beneficial Reuse of Biosolids: How Important is Public Perception? In *Biosolids Public Acceptance-Digest,* Alexandria, VA: Water Environment Federation.

Draman, Grace A. 1995. Public Perception is Key to Biosolids Acceptance. *BioCycle*, 36(9), 82-83.

Egigian-Nichols, C. and C. Lekven. 1998. All New Biosolids Fact Sheets – 26 Flavors from Sacramento. In *12th Annual Residuals and Biosolids Management Conference*, Bellevue, WA: Water Environment Federation. 535-542.

Ehrmann, John R., and Barbara L. Stinson. 1999. Joint Fact-Finding and the Use of Technical Experts. In Lawrence Susskind, Sarah McKearnan, and Jennifer Thomas-Larmer (eds.), *The Consensus Building Handbook: A Comprehensive Guide to Reaching Agreement.* Thousand Oaks, CA: Sage Publications.

Englebow, Jill. 1992. Co-composting Controversy: County Recycler Sees Spot on Team For Sludge. *The Maui News*. May 31.

Environmental Working Group. 1998(a). Dumping Sewage Sludge on Organic Farms? Why USDA Should Just Say No. April 30. http://www.enviroweb.org/issues/sludge.

Environmental Working Group. 1998(b). Toxic Wastes "Recycled" as Fertilizer Threaten U.S. Farms and Food Supply: Dioxin, Lead, Mercury Spread on Crops as States Scramble to Protect Public Health. http://www.ewg.org/pub/home/reports/factoryfarming/fertpress.html

European Union. 2000. Working Document on Sludge, 3rd Draft. Brussels. April 27.

Evanylo, Greg. 1998. *Biosolids Land Use Ordinances Gaining Favor.* Virginia Cooperative Extension. January. http://www.ext.vt.edu/news/periodicals/cses/1998-01/1998-01-05.html

Evanylo, G.K. 1999. *Agricultural Land Application of Biosolids in Virginia: Risks and Concerns.* Virginia Cooperative Extension. Publication 452-304. Available at http://www.ext.vt.edu/pubs/compost/452-304/452-304.html

Farrell, Molly. 2001. Site Monitoring and Odor Control at Co-composting Facility. *BioCycle.* July.

Fischhoff, B. 1999. Why (Cancer) Risk Communication Can be Hard. *Journal of the National Cancer Institute Monographs*, 25, 7-13.

Fitzcharles, Margaret. 2000. Norco Woman Fights DEP Over Spread of Sewer Sludge. *The Mercury,* PA, September 7.

Fondahl, Lauren. 1999. Biosolids Management in the Western Region. *BioCycle*. July.

Forste, J.B. 1993. Gaining Public Acceptance for Biosolids. *Sewage Sludge: Land Utilization and the Environment*, Bloomington, MN: Soil Science Department and USDA-ARS, University of Minnesota. August 11-13.

Forste, Jane. 1995. Building a Biosolids Program: Elements of Success. *The Biosolids Trilogy*. Silverdale, WA. Northwest Biosolids Management Association 7th Annual Biosolids Management Conference.

Forste, J. 2000. Bridging the Communications Gap: The EMS Approach. *14th Annual Residuals And Biosolids Management Conference,* Boston, MA.

Forste, J.B. and P.S. Machno. 1994. Public Acceptance of Biosolids--What's In a Name? In *The Management of Water and Wastewater Solids for the 21st Century: A Global Perspective*, Washington, D.C.: Water Environment Federation. 9-45–9-55.

Fraser, Patrick. 2000. Deadly Dirt? *WSVN TV News,* FL. November 14. Sunbeam Television Corporation.

Frederick Schneiders Research. 1998. Survey results summary presented to the Water Environment Federation.

Freeman, D. 1995. Controlling Odors: Don't Let Them Get The Drift. *World Wastes* 38(9):40. September.

Freudenberg, W.F. 1996. Risky Thinking: Irrational Fears About Risk and Society. *Annals of the American Academy of Political and Social Science*, 545, 44-63.

Gardner, G. 1997. *Recycling Organic Waste: From Urban Pollutant To Farm Resource.* Worldwatch paper #135. Washington, DC: Worldwatch Institute.

Gardner, G. 1998. Recycling Human Waste: Fertile Ground Or Toxic Legacy? *World Watch* 11(1):28-34. January/February.

General Physics Corp. Environmental Services Division. 1991. Data results: Revised bioaerosol monitoring program for the Washington Suburban Sanitary Commission. Montgomery County Regional Composting Facility. Silver Spring, MD.

Gilbert, Steve and Sue Hennig. 1996. Biosolids Long-Range Planning: Changing the Paradigm. *A Bridge to Sustainability*. Chelan, WA. Northwest Biosolids Management Association's 8th Annual Biosolids Management Conference.

Goldfarb, W., U. Krogmann, and C. Hopkins. 1999. Unsafe Sewage Sludge or Beneficial Biosolids?: Liability, Planning, and Management Issues Regarding the Land Application of Sewage Treatment Residuals. *Boston College Environmental Affairs Law Review*. 26:687.

Goldstein, Nora. 2000. National Overview of Biosolids Management. *BioCycle,* December.

Gray, George, Kim Thompson, and Pamela Williams. 2002. Risks of Conventional and Organic Produce: Public and Scientific Perceptions. Harvard Center for Risk Analysis: http://www.hcra.harvard.edu/foodresearch.html#tradeoffs

Greater Vancouver Regional District. 2000. *Issues on Biosolids: Report on Focus Groups and Triads.* December 4.

Greenwood, Gail. 1997. How to Get the News Hounds to Chase you Without Getting Bit. *Illuminating the Future.* Vancouver, WA. Northwest Biosolids Management Association's 9th Annual Biosolids Management Conference.

Haag, Ed. 1992(a). Just Say No. *Dairy Today.* March, 82-83.

Haag, Ed. 1992(b). Sludge Under Suspicion. *Farm Journal.* 116 (6): 16-19.

Hadeed, Sam. 1999. Developing An Effective & Sustainable NBP Communications Strategy, presentation to *Reinvesting in Success*, a meeting of biosolids stakeholders and regional associations, Charlotte, NC.

Hale, Robert. 2001. *Nature*. 412:140-141.

Hansen, Renee. 1995. Hitting the Fan. *Snoqualmie Valley Reporter*. January 18.

Harding, Earle C. 2001. Laymanization: An Engineer's Guide to Public Relations. *Water Environment and Technology.* P. 45-48. April.

Harrison, Ellen Z. 2000. Desired Outcomes of Hearing, Comments to the U.S. House Committee on Science, March 22.

Harrison, Ellen Z., Murray McBride and David Bouldin. 1999. *The Case for Caution: Recommendations for Land Application of Sewage Sludges and an Appraisal of the U.S. EPA's Part 503 Sludge Rules*. Cornell Waste Management Institute: August 1997 and February 1999. Available at http://www.cfe.cornell.edu/wmi/PDFS/LandApp.pdf.

Hartley, Troy W. 2003. *Public Perception & Participation in Water Reuse*. Alexandria, VA: Water Environment Research Foundation, National Water Research Institute (NWRI), American Water Works Association Research Foundation (AWWARF), WateReuse Foundation. August.

Hassell, Leroy R., Sr. 2001. Virginia Supreme Court Opinion January 12. Reuben L. Blanton, et al. v. Amelia County, et al. Record No. 000277, from the Circuit Court of Amelia County.

Hazen & Sawyer. 1995. Task 15.6 Final Public Participation Report, Volume 1. Concerning the New York biosolids management planning process.

Hegedus, Mike. 2000. The Marketing of Nature: The Nature of Marketing. Cleveland, OH: U.S. Composting Council's 10th Annual Conference.

Hennig, Sue. 1995. Biosolids Compost: A Collaborative Solution for Biosolids Public-Acceptance. *The Biosolids Trilogy*. Silverdale, WA. Northwest Biosolids Management Association 7th Annual Biosolids Management Conference.

Henry, Tom. 2002. Skeptics critical of use of sludge on farmlands. *Toledo Blade.* Toledo, OH. November 25.

Herkert, J.R. 1994. Ethical Risk Assessment: Valuing Public Perceptions. *IEEE Technology and Society Magazine*, 79, 4-10.

Hirshorn, Susan. 1992. Human Fertilizer Goes Out To Pasture. *Canadian Living.* February.

Hirst, E. 1984. Household Energy Conservation: A Review of the Federal Residential Conservation Service. *Public Administration Review,* 44, 421-430.

Hodson, C.O. 1996. Biosolids Management: Beneficial Use Comes of Age. *Pollution Engineering* 28(13):38-41. December.

Hunt, Patricia, and David W. Oerke. 1994. Institutions, Attitudes, and Opinions Impacting Acceptance of Biosolids Land Application Programs in Northeastern Colorado. In *Biosolids Public Acceptance-Digest,* Alexandria, VA: Water Environment Federation, 1996.

Hyde, James. 2002. Communicating About Risk During Chaos Events. Presentation to the New England Water Environment Annual Conference. January 28.

Inlow, Shawn K. 2000. George, Tackett Assail Use of Sludge. *The Progress,* July 17.

International Association for Public Participation. 2000. *IAP2 Public Participation Toolbox.* From the Internet site http://www.iap2.org.

Jefferson County, AL. 2000. Beltona Land Application Program. Videotape.

Johnson, Dawn. 1997. Golf club expansion begins. *The Similkameen Spotlight.* July 23.

Jones, Robert E., and Riley E. Dunlap. 1992. The Social Bases of Environmental Concern: Have They Changed Over Time? *Rural Sociology*, 57(2), 28-47.

Kamrin, Michael; Dolores J. Katz; and Martha L. Walter. 1995. *Reporting on Risk: A Journalist's Handbook on Environmental Risk Assessment.* Ann Arbor, MI: Michigan Sea Grant Program and the Foundation for American Communications. A shorter version of this handbook (posted in 2000) is available from FACSNET at http://www.facsnet.org/tools/ref_tutor/risk/index.php3.

Kelsey, T.W. and L. Singletary. 1996. When Composting Meets Suburbia. *BioCycle* 37(7):66-67.

Keremeos Review, The. 1999(a). Biosolids Answers Needed. February 25.

Keremeos Review, The. 1999(b). Pollution Would Be Hard To Prove. February 11.

Keremeos Review, The. 2000. Permit Given in '99. October 12. #

Kern County Biosolids Ordinance Adoption Process. 2002. http://www.co.kern.ca.us/rma/biosolids2002.asp

Kimantas, John. 2000. *Waste Not, Want Not*. September 11.

King County, Washington. 2000. Biosolids recycling program. http://dnr.metrokc.gov/WTD/biosolids/BMP1.html.

Kirk, Ken and Albert Gray. 2000. Letter to Marilyn Fingerhut, Ph.D., Chief of Staff, National Institute for Occupational Safety and Health (NIOSH), October 31. American Metropolitan Sewerage Agencies and Water Environment Federation response to NIOSH HID #10 regarding Class B biosolids workers.

Kliewer, T. 1999. Firm's Bid To Spread Treated Sludge On Farm Worries Waller County. *The Houston Chronicle, Star Edition*. 13. May 17.

Krauss, G.D. and A.L. Page. 1997. Wastewater, Sludge and Food Crops. *BioCycle*. 38(2):74-82. February.

Krogmann, Uta and Ellen Z. Harrison. 1998. Understanding the difference: Why European and U.S. sludge standards differ. Paper presented at AAAS Conference, February, 1998, Philadelphia. http://www.cfe.cornell.edu/wmi/Sludge/AAAS/Overview.html).

Kunkle, Fredrick. 2001. Sludge-Spreading Raising Concern Over Health Fears. *The Washington Post.* August 23.

Lang, M. 2000. The Biosolids Message. In *Proceedings from 14th Annual Residuals and Biosolids Management Conference*, Boston, MA: Water Environment Federation. February 27-29.

Lang, M.E. and R.A. Jager. 1996. The Impacts of the Public's Perception on Land Application Programs Implemented Under the 503 Regulations. In *10th Annual Residuals and Biosolids Management Conference: 10 Years of Progress and a Look Toward the Future*, Denver, CO. 4-31–4-35. August.

Lawson, Lisa. 1998. Cawston opposes biosolids. *The Similkameen Spotlight.* October 6.

League of Women Voters Education Fund. 1976. *Municipal Sludge: What Shall We Do With It?* Publication No. 627.

Leeder, Jessica. 2002. Human Fertilizer Poses Cancer Risk: Study. *National Post.* July 31.

Lees, P.S.J. and M.S. Tockman. 1987. Evaluation of possible public health impact of WSSC Site II sewage sludge composting operations. Johns Hopkins University School of Hygiene and Public Health, Baltimore, MD.

Lester, James P. and Ann O'M. Bowman, eds. 1983. *The Politics of Hazardous Waste Management.* Duke University Press, Durham, North Carolina.

Lewis, David L. 2000(b). Political Science at its Worst: The 503 Sludge Rule: Why EPA Regulations Have Everything to do with Elections and Nothing to do with Protecting Public Health and the Environment. http://www.whistleblowers.org/statements.htm#state2.

Lewis, D.L., S. Shepherd, D.K. Gattie, S. Sanchez, and M. Novak. 2000. Enhanced Susceptibility to Infection from Exposure to Gases Emitted by Sewage Sludge: A Case Study. In

Biosolids Management in the 21st Century Conference, College Park, MD. 152-159. Available at http://members.aol.com/LewisDaveL/Proceedings.htm.

Lewis, David and David Gattie. 2002. Pathogen Risks from Applying Sewage Sludge to Land. *Environmental Science and Technology*, July 1, pp. 287-293.

Lewis, David, David Gattie, Marc Novak, Susan Sanchez, and Charles Pumphrey. 2002. Interactions of Pathogens and Irritant Chemicals in Land-Applied Sewage Sludges (Biosolids). *BMC Public Health* 2:11, June 28.

Libby, John. 1995. Brain Viruses on the Prairie! A Biosolids Siting Tragicomedy. *The Biosolids Trilogy*. Silverdale, WA. Northwest Biosolids Management Association 7th Annual Biosolids Management Conference.

Lindsay, B.E., H. Zhou, and J.M. Halstead. 2000. Factors Influencing Resident Attitudes Regarding The Land Application Of Biosolids. *American Journal of Alternative Agriculture* 15(2):88-95.

Lloyd, Jillian. 1998. Special Report: A Trooper, A Dump, and a Tale of Doubt. *The Christian Science Monitor*. June 10.

Logan, T.J. 1995. Gaining Public Acceptance for Beneficial Use of Biosolids. *BioCycle*. 36(12): 61-64.

Logan, T.J. 1997. Balancing Benefits and Risks In Biosolids. *BioCycle* 38(11):52-57.

Logsdon, G. 1987. Cultural Attitudes and Waste Disposal. *BioCycle,* March.

Logsdon, G. 1992. How Does Society Learn About Sludge Safety? *BioCycle* 33(5):68-70. May.

Logsdon, G. 1995 *The Contrary Farmer*. CA: Real Goods.

Loken, Lorraine. 1994. A National Strategy for Public Acceptance of Biosolids. *Conveying the Message*. Chelan, WA. Northwest Biosolids Management Association's 6th Annual Biosolids Management Conference.

Lopez, Elizabeth. 1995. Discovery Park Sludge Treatment To Help Restore Native Plant Life. *The Seattle Times*. August 4.

Lue-Hing, Cecil. 2001. Risks From Toxicants and Pathogens in Biosolids Fertilizers. Comments to the National Research Council, Public Meeting, March 14.

Machno, P.S. 1996. Biosolids 2000: Are We Letting Our Guard Down on Public Acceptance? In *10th Annual Residuals and Biosolids Management Conference: 10 Years of Progress and a Look Toward the Future*, Denver, CO. 4-21–4-27. August.

Machno, Peter. 2000. National Biosolids Partnership Environmental Management System Overview. *14th Annual Residuals and Biosolids Management Conference,* Boston, MA. Alexandria, VA: Water Environment Federation.

Machno, P.S. and J. Forste. 1997(a). Biosolids 2000, Public Acceptance of Biosolids Recycling. In *Water Residuals and Biosolids Management Approaching the Year 2000*, Philadelphia, PA: WEF/AWWA Joint Conference. 4-1–4-9.

Maddocks, Susan. 1995. Fear's Scent Remains Over Sludge. *Portsmouth Herald,* NH. December.

Mahin, Tom, Richard Pope, and Charles McGinley. 2000, When Is Smell a Nuisance? An Overview of Different Approaches Taken Around the World In Setting Odor-Control Measures. *Water Environment and Technology,* May. Alexandria, VA: Water Environment Federation.

Mahr, Joe. 2000. Cash-Strapped Farmers Leery of Sludge Warnings. *The Blade,* Toledo, OH. www.toledoblade.com.

Maryland Department of the Environment. 1998. *Sewage Sludge Utilization in Maryland*, fact sheet.

Massad, Jason. 2002. Growing Question: Are Biosolids Safe? Sludge Concern Spreads. *Vacaville Reporter.* California. August 24.

Matthews, Peter. 1997. Transatlantic Comparison of Biosolids Practices. Anglian Water Services, UK.

Maui News, *The.* 1998. EPA Gives Thumbs Up to Co-Composting Facility, Calls Product Exceptional. November 2.

Mayer, Roger. 1999. Mayer Defends Biosolids. *The Keremeos Review.* February 18.

Mazmanian, Daniel and David Morell. The NIMBY Syndrome: Facility Siting and the Failure of Democratic Discourse. In Norman Vig and Michael Draft, eds., *Environmental Policy in the 90s.* CQ Press, Washington, D.C.

McAvoy, Gregory E. 1999. *Controlling Technocracy: Citizen Rationality and the NIMBY Syndrome.* Georgetown University Press, Washington, D.C.

McBride, M.B. 1995. Toxic Metal Accumulation from Agricultural Use of Sludge: Are USEPA Regulations Protective? *Journal of Environmental Quality*, 24:5-18.

McCance, M. 1998. Board Bans Sludge Use In County. *Richmond Suburban Newspapers*. June 20-26.

McConnell, D. Hew. 1994. GVRD takes issue with comments on Delta sewage plant. *South Delta Today.* December 11.

McDaniels, T.L., L.J. Axelrod, N.S., Cavanagh, and P. Slovic. 1997. Perception of Ecological Risk to Water Environments. *Risk Analysis*, 17(3), 341-51.

McDougall, Ruth, Michael D. Van Ham, and Mary Jane Douglas. 2002. *Best Management Practices Guidelines for the Land Application of Managed Organic Matter in British Columbia.*

McKenzie-Mohr, Doug. 1996. *Promoting a Sustainable Future: An Introduction to Community-Based Social Marketing*. National Round Table on the Environment and the Economy, Ontario (CN) Ministry of Environment and Energy.

McKinney, D.E., T. Immerso, S. Fangmann, and C. Koch. 1990. NIMBY and the Ocean Dumping Ban Act of 1988. In *The Water Pollution Control Federation Specialty Conference Series, The Status of Municipal Sludge Management for the 1990's*. New Orleans, LA: The Water Pollution Control Federation. 90-92.

Meadows, Donella. 1997. In the Great Sludge War, Spreading Comes Out Ahead: Done Well, It Makes Ecological Sense. *Concord Monitor.* February.

Mendenhall, T.C. 1991(a). A Municipal Perspective. *National Association of Professional Environmental Communicators* 2(4). December.

Mendenhall, T.C. 1991(b). Strategic Planning for Residuals (Sludge) Management. Presentation at BioCycle Southeast Conference, Raleigh, NC. November 20.

Mendenhall, T.C. 1991(c). Environmental Communication: Strategy and Tactics. *AWWA/WPCF Joint Residuals Management Conference "Residuals Management After 1991."* Durham, NC: American Water Works Association and Water Pollution Control Federation. August 11-14.

Mendenhall, T.C. 1992. Public Acceptance of Biosolids--Biosolids Managers' Perspective. In *Biosolids Public Acceptance-Digest,* Alexandria, VA: Water Environment Federation.

Merrimack, NH. 2000. *Merrimack: The Town and the River.* Videotape. 38 minutes.

Merriman, Ed. 1991. Farmers, Public Warned of Sludge Danger. *The Capital Press,* WA. July 19:1.

Metropolitan Water Reclamation District of Greater Chicago, 1995. Website description of biosolids program. http://www.mwrdgc.dst.il.us

Metropolitan Water Reclamation District of Greater Chicago. 1999(a). Land Application of Sewage Sludge: Papers and Publications by the R&D Department, and Research Funded by the District - A Bibliography, 1968 -1998, Rpt. #99-11.

Metropolitan Water Reclamation District of Greater Chicago. 1999(b). Improvements in the Quality of Sewage Sludge at the Metropolitan Water Reclamation District of Greater Chicago, Rpt. #99-20.

Metropolitan Water Reclamation District of Greater Chicago. 1999(c). Effect of Time After Cessation of Biosolids Applications on the Concentration of Cadmium, Copper, Nickel, and Zinc in Soil, Leaves and Grain of Corn, Rpt. #99-23.

Metropolitan Water Reclamation District of Greater Chicago. 1999(d). Environmental Protection System Report for Fulton County, Illinois, May 1999, Rpt #99-18.

Miller, G.M., B.A. Janonis, and J.A. Billica. 1996. Soft Engineering: Management of Public Concerns in a Potentially Controversial Biosolids Project. In *10th Annual Residuals and Biosolids Management Conference: 10 Years of Progress and a Look Toward the Future*, Denver, CO. 4-15–4-20. August.

Milwaukee Metropolitan Sewerage District (MMSD), 1995. Website: http://www.milorganite.com

Minnich, Jerry, Marjorie Hunt, and the Editors of Organic Gardening Magazine.1979. *The Rodale Guide to Composting.* Emmaus, PA: Rodale Press.

Missouri Extension. 1996. Safety and Benefits of Biosolids. http://muextension.missouri.edu/xplor/waterq/wq0427.htm.

Mittelstaedt, Martin. 1999(a). Guidelines Would Likely Increase. *The Globe and Mail.* January 20.

Mittelstaedt, Martin. 1999(b). Sewage Sludge Gaining Acceptance As Farm Fertilizer. *The Globe and Mail.* January 20.

Montague, P. 1999. Excrement Happens. *The Ecologist* 29(4):267-269. July.

Motavalli, Jim. 1999. The 20/20 Vision Thing: The Green Group for Busy People. *E Magazine.* March-April.

Mulvihill, Keith. 2001. Long-Lasting Pollutant Found in Fertilizer, Fish. *Reuters.,* July 11.

National Biosolids Partnership. 1999(a). Code of Good Practice. *1999-2000 Annual Report.* Alexandria, VA: National Biosolids Partnership.

National Biosolids Partnership. 2000(a). *1999 – 2000 Annual Report.*

National Biosolids Partnership. 2000(b). *National Manual of Good Practice for Biosolids.* Alexandria, VA.

National Biosolids Partnership. 2001. *2000-2001 Annual Report.* Alexandria, VA: National Biosolids Partnership. www.biosolids.org.

National Biosolids Partnership. 2002(a). *Third Party Verification Auditor Guidance.* April, 2002.

National Biosolids Partnership. 2002(b). *Elements of an Environmental Management System (EMS) for Biosolids,* final interim draft, May 1, 2002.

National Institute for Occupational Safety and Health (NIOSH). 2000. *Workers Exposed to Class B Biosolids.* Hazard ID #10. Morgantown, WV: NIOSH.

National Institute for Occupational Safety and Health (NIOSH). 2002. *Guidance For Controlling Potential Risks To Workers Exposed to Class B Biosolids.* Replaced the NIOSH Hazard ID #10 (see above).

National Research Council, 1996. *Understanding Risk: Informing Decisions in a Democratic Society.* Washington, D.C.: National Academy of Sciences.

National Research Council. 2001. Study of Public Participation in Environmental Assessment and Decision Making. Summary Notes from Planning Meeting. July 20.

National Research Council. 2002. *Biosolids Applied to Land: Advancing Standards and Practice.* Washington, D.C.: National Academy Press. Prepublication Copy, July.

National Research Council.1996. *Use of Reclaimed Water and Sludge in Food Crop Production,* Washington D.C.: National Academy Press.

National Sludge Alliance. 1997(a). NSA Fact Sheet 104: EPA's Reckless Endangerment of Public Health. February 10. http://www.enviroweb.org/issues/sludge/nsa/nsa104.html.

National Sludge Alliance. 1997(b). NSA Fact Sheet 107: The Sludge Gets Deeper. March 10. http://www.enviroweb.org/issues/sludge/

National Sludge Alliance. 1997(c). Public Facts 113: Is it Toxic Sludge or Cow Manure Poisoning Our Food Supply. May 7. http://www.enviroweb.org/issues/sludge/.

National Sludge Alliance. 1997(d). Caution: EPA Scientist at Work. NSA Fact Sheet 111. April 16.

National Sludge Alliance. 1997(e). EPA's Reckless Endangerment of Public Health. February 10. http://www.enviroweb.org/issues/sludge/nsa/nsa104.html.

National Sludge Alliance. 1997(f). Is It Toxic Sludge or Cow Manure Poisoning Our Food Supply? May 7. http://www.enviroweb.org/issues/sludge.

National Sludge Alliance. 1997(g). The Sludge Gets Deeper. March 10 http://www.enviroweb.org/issues/sludge.

National Small Flows Clearinghouse. 1998. Managing Biosolids in Small Communities. *Pipeline.* Vol 9, No. 4, Fall 1998.

National Whistleblower Center. 2002. Scientific Freedom Under Attack: An Analysis of the Synagro Technologies, Inc. "White Paper" Regarding Dr. David Lewis. January 24. (Available at: http://www.whistleblowers.org/sludgereplyfinal.htm)

National Wind Coordinating Committee. 2002. *Permitting of Wind Energy Facilities, A Handbook*. National Wind Coordinating Committee: http://www.nationalwind.org . August.

Nature's Blend Organic Fertilizers. 2000. Advertising brochure. Warren, OH.

Nesmith, Jeff. 1996. Turmoil at EPA. *The Atlanta Journal-Constitution, Saturday Reader.* July 27.

New England Biosolids and Residuals Association (NEBRA). 1999(a). Questions and Answers. http://www.nebiosolids.org

New England Biosolids and Residuals Association. 1999(b). Communications System for the National Biosolids Partnership and Regional/Local Groups: Suggestions from the Reinvesting in Success Workshop, Charlotte, NC.

New England Biosolids and Residuals Association (NEBRA). 2001. Were Biosolids Involved in the Death of a Greenland, New Hampshire Man? Information Update. Available at http://www.nebiosolids.org/scienceof.html.

New England Interstate Water Pollution Control Commission. 1994. Sludge, a.k.a. Biosolids! *Water Connection,* 11(2), Fall.

New England Interstate Water Pollution Control Commission. 1995. *Land Application of Biosolids*. Fact Sheet prepared by NEIWPCC's Residuals Workgroup. October.

New Hampshire Comparative Risk Project. 1997. Report of Ranked Environmental Risks in New Hampshire, Concord, NH.

New Hampshire Water Supply and Pollution Control Commission. 1978. Land Application of Wastewater Sludge at Somersworth, NH. November.

New Hampshire, University of. 1998. Landspreading of Sludge in New Hampshire, Report of the UNH Sludge Task Force.

New York City Department of Environmental Protection, Biosolids Citizen Advisory Committee. 1998-2000. Minutes of Meetings.

New York State Department of Environmental Conservation.1997. *A Technical Review of 'The Case for Caution,'* document finalized and made available as part of rulemaking process in 2002.

Norman, Roni. 1995. Reader objects to Nutrifor Treatments. *The Similkameen Spotlight.* September 13.

Northwest Biosolids Management Association (NBMA). 1998(a). *Biosolids Recycling: Recognizing a Resource.* Fact Sheets, video.

Northwest Biosolids Management Association. 1997-1998. *Literature Reviews*.

Northwest Biosolids Management Association. 1999. *The Results of the Marketing Assessment for Biosolids.* April.

O'Dette, R.G. 1994. The Beneficial Use of Biosolids: We're Getting There! In *The Management of Water and Wastewater Solids for the 21st Century: A Global Perspective*, Washington, D.C.: Water Environment Federation. 9-1–9-12.

Oerke, D.W., J.H. Rickermann, F. Bebler, and P. Heppler. 1998. What Happened When Neighbors Moved Next Door – A Successful Approach to Public Involvement in Biosolids Management Decisions in Boulder, Colorado. In *12th Annual Residuals and Biosolids Management Conference*, Bellevue, WA: Water Environment Federation. 555-561.

Orlando, L. 1997. The Sewage Scam – Should Sludge Fertilize Your Vegetables? *Dollars and Sense* 211:34-37. May/June. http://www.riles.org/paper2.htm.

Orlando, Laura. 1999(a). Something Stinks in the EPA. ReSource Institute for Low Entropy Systems http://www.riles.org/musings23.htm. September 24.

Orlando, Laura. 1999(b). Toxic Avengers: The EPA is Pushing Hazardous Sludge as Fertilizer. The Locals are Fighting Back. *In These Times*. February 21. http://www.riles.org/library.htm.

Outwater, Alice. 1996. *Water, A Natural History.* New York: Basic Books.

Ozawa, Connie P. and Lawrence E. Susskind. 1984. Mediating Science-Intensive Public Policy Disputes. Presented at the annual meeting of the Association for Public Policy Analysis and Management, October 18-20.

Pearce, Robert. 1997. www.nosludge.org. Palmdale, CA.

Pearlman, Stephen and Steve Frank. 2001. Winning Trust Through Open, Honest Communication Helps Reduce Neighbors' Fears of Biosolids. *15th Annual Residuals and Biosolids Management Conference, San Diego, CA.* Water Environment Federation conference.

Pearlman, Stephen, Duane E. Humble, and Stephan D. Frank. 2000. Independent Biosolids Monitoring Plan Paves the Way for Cooperation Between Skeptical Neighbors and Wastewater Agency. *14th Annual Residuals and Biosolids Management Conference, Boston, MA.* Water Environment Federation conference.

Pellow, D.N. 1999. Framing Emerging Environmental Movement Tactics: Mobilizing Consensus, Demobilizing Conflict. *Sociological Forum*, 14(4), 659-683.

Pennsylvania Department of Environmental Protection. 2000. Regulation and Beneficial Use of Biosolids. Internet website: http://www.dep.state.pa.us/dep/biosolids/toc.htm.

Pennsylvania Environmental Network. 1999?. The Power to Protect! Flyer. Available at http://www.penweb.org.

Pennsylvania Environmental Network. 2001. Why We Are Opposed to the Land Application of Sewage Sludge. Received via email. July. Reprinted in Appendix B.

People Against Power's Sludge (PAPS). 1999. Letter to local citizens regarding land application of biosolids permitting in Rush Township, PA.

Peters, R.G., V.T. Covello, and D.B. McCallum. 1997. The Determinants of Trust and Credibility in Environmental Risk Communication: An Empirical Study. *Risk Analysis*, 17(1), 43-54.

Pick, C. 1996. Bouncing Back From a Public Nuisance Setback. *BioCycle* 37(9):58-61.

Pooley, Julie Ann and Moira O'Connor. 2000. Environmental Education and Attitudes: Emotions and Beliefs Are What is Needed. *Environment and Behavior*, 32(5), 711-724.

Powell, Douglas. 1996. An Introduction to Risk Communication and the Perception of Risk. Unpublished manuscript, University of Guelph, Ontario, Canada.

Powell-Tate. 1993. *Communications Plan on Biosolids.* Prepared for the Water Environment Federation.

PR Watch.1995. A R.O.S.E. by Any Other Name… *PR Watch,* 3rd Quarter.

PR Watch. 1999(b). Sandman's Cagey Tactics (letter to PR Watch) www.prwatch.org. **#3.** PR Watch and readers attack Peter Sandman's model of public outrage; "outrage can be good; hold on to it."

Providence Journal. 2002. Internal Review Faults EPA Over Safety of Recycled Sewage. February 7.

Racey, P.K. 1991. Plain Talk – Putting the Risks of Land Application Into Perspective for the Public. *AWWA/WPCF Joint Residuals Management Conference "Residuals Management After 1991."* Durham, NC: American Water Works Association and Water Pollution Control Federation. August 11-14.

Rachel's Environment & Health Weekly. 1997. Sewage Sludge, A New U.S. Waste Policy Emerges, Parts 1 and 2, August 21 and 28.

Rainey, M. 1998. Working Out Approval Of Biosolids Land Application. *BioCycle* 39(5):66-69.

Rampton, S. 1998. Let Them Eat Nutri-cake: Merriam-Webster Thinks Our "Biosolids" Don't Stink. *Harper's Magazine* 297(1782):48. November.

Randall, Tom. 2000. A Death in the Night, Yet EPA Maintains Breathing Dried Human Waste is Safe. *Environment News*. June.

Reardon, Kate. 1998. Public Works Scientist Instrumental In Getting Word Into Dictionary. *The Herald*. July 25.

Reed, Chris. 2000. Kern Tiring of O.C. Sludge. *The Orange County Register, Sunday*. July 16.

Rembert, T.C. 1998. Strategic Retreat on Organics. *E Magazine: The Environmental Magazine* 9(4):27. July/August.

Renn, Ortwin, Thomas Webler, and Branden B. Johnson, Summer 1991, "Public Participation in Hazard Management: The Use of Citizen Panels in the U.S." *RISK—Issues in Health & Safety*, 197-226.

Renn, Ortwin, Thomas Webler, and Peter Wiedemann. 1995. *Fairness and Competence in Citizen Participation: Evaluating Models for Environmental Discourse*. Boston: Kluwer Academic Publishers.

Rich, Richard C., Michael Edelstein, William K. Hallman, and Abraham H. Wandersman, 1995, Citizen Participation and Empowerment: The Case of Local Environmental Hazards. *American Journal of Community Psychology*, 23(5):657-.

Rockefeller, Abby A. 1999. Civilization & Sludge: Notes on the History of the Management of Human Excreta. *Current World Leaders*, Vol. 39, No. 6. Also available at www.riles.org/paper1a.htm.

Rocky Mountain Water Environment Association. *Recycling Completes the Cycle*. Videotape.

Rocky Mountain Water Environment Association. 1995. *Guide to Dealing With the Media*.

Rocky Mountain Water Environment Association. 1996. Biosolids Factsheets. RMWEA Biosolids Committee.

Rodgers, S. 1994. The Role of Local Government in Land Application of Biosolids Programs. In *The Management of Water and Wastewater Solids for the 21st Century: A Global Perspective*, Washington, D.C.: Water Environment Federation. 9-73–9-80.

Ropeik, David and George Gray. 2002. Risk. Harvard Center for Risk Analysis.

Rosenbaum, Walter. 1983. The Politics of Public Participation in Hazardous Waste Management. *The Politics of Hazardous Waste Management*, Duke University Press, Durham, NC.

Rosenfeld, Paul. 2000. The Answer is Blowing in the Wind: Options for Odor Control. *Storms on the Horizon*. Ocean Shores, WA. Northwest Biosolids Management Association 11th Annual Biosolids Management Conference.

Ross & Associates Environmental Consulting. 2000. *Recommendation to the National Biosolids Partnership on a Biosolids EMS Third Party Verification Program*. Alexandria, VA: National Biosolids Partnership.

Rowe, G. and L.J. Frewer. 2000. Public Participation Methods: A Framework for Evaluation. *Science, Technology & Human Values*, 25(1), 3-29.

Rubens, Jim. 2000. NH Sludge Regulations Are Lax. *Valley News,* August 16.

Rubin, Dr. Alan B. 1998. Statement before the Kern County Biosolids Ordinance Advisory Committee in Bakersfield California. December 1.

Rynk, R., M. van de Kamp, G.B. Willson, M.E. Singley, T.L. Richard, J.J. Kolega, F.R. Gouin, L. Laliberty, Jr., D. Kay, D.W. Murphy, H.A.J. Hoitink, and W.F. Brinton. 1992. *On-farm Composting Handbook*. Ithaca, NY: Northeast Regional Agricultural Engineering Service.

Sacramento Regional Wastewater Treatment Plant—Biosolids Program. 1999. Biosolids Fact Sheets.

Sacramento Superior Court. 2001. Sacramento Superior Court WRIT: Central Delta Water Agency, et. al. vs. State Water Resources Control Board. July 20.

Sandman, P.M. 1987(a). Communicating Risks: Some Basics. *Health and Environment Digest,* Vol. 1, No. 11, 3-4.

Sandman, P.M. 1987(b). Risk Communication: Facing Public Outrage. *EPA Journal,* Vol. 13, No. 9, 21-22.

Sandman. (1993). *Responding to Community Outrage: Strategies for Effective Risk Communication.* Fairfax: American Industrial Hygiene Association.

Sandman, P.M., P.M. Miller, B.B. Johnson, and N.D. Weinstein. 1993. Agency Communication, Community Outrage, and Perception of Risk: Three Simulation Experiments. *Risk Analysis*, 13(6), 585-97.

Sandman, P.M. 2000. Dealing With Outrage: A Key Communication Tool for Biosolids Professionals, Handbook. Workshop at *14th Annual Residuals and Biosolids Management Specialty Conference.* Water Environment Federation.

Sarber, K. 1994. How to Strategize for Successful Project Development. *BioCycle* 35(4):32-35.

Sasser, Larry. 1993. Exceptional Quality Biosolids: The Key to Public Acceptance. *Entering a New Era.* Alderbrook, WA. Northwest Biosolids Management Association Annual Biosolids Management Conference. 12.

Scharp, Michael. 2000. Why Colorado Farmers Want New York City Biosolids. *BioCycle.* July, 71-80.

Scharp, Michael. 2001. The Nuts and Bolts of a Successful Land Application Project: NYC Biosolids in Colorado. WEF Joint Residuals and Biosolids Management Conference, San Diego, CA.

Scherer, Clifford W. 1993(b). *Public Participation: Risk and Waste Management.* National Extension Compost Utilization Conference, Minneapolis, MN.

Schiffman, Susan S., J.M. Walker, P. Dalton, T. S. Long, J. H. Raymer, D. Schusterman, and C. M. Williams. 2000. Potential Health Effects of Odor from Animal Operations, Wastewater Treatment, and Recycling of Byproducts. *Journal of Agromedicine,* 7 (1).

Scruton, D. 1991. Memorandum to George Dunsmore, Commissioner of Agriculture regarding "Ruane Farm, No. Clarendon, VT." Vermont Department of Agriculture. June 27.

Sewage Sludge Homepage. 2000. Sludge on Your Supper Table. http://www.enviroweb.org/issues/sludge/

Shaheen, Jeanne. 2000. A Proclamation: Biosolids Recycling Day, May 16, 2000. Concord, NH.

Shields, Helane. 1997. Shields Urges Voters to Petition for Articles to Ban the Use of Sludge to be Placed on Town Warrants. *Granite State News,* January 22.

Shields, Helane. 1998. Sludge on Farmland Case Studies. www.websida.com/danger/case.htm. January.

Sierra Club, New Hampshire Chapter. 1999. Sewage Sludge in New Hampshire: Questions and Answers. Concord, NH: NH Sierra Club.

Sierra Club, New Hampshire Chapter. 2000. *Policy Driven Water Pollution: The Dangers of Using Sludge to Reclaim Gravel Pits in New Hampshire.* Concord, NH: Sierra Club/NH.

Sierra Club, Pennsylvania Chapter. 2000(a). Sewage Sludge Hits the Fan in Centre County, A Public Letter from Len Martin and PAPS: People Against Power Sludge. http://pennsylvania.sierraclub.org/ moshannon/fctryfrm.htm. 2000.

Sierra Club, Pennsylvania Chapter. 2000(b). Statement regarding *E. coli* contamination from manure at Walkerton, Ontario. http://pennsylvania.sierraclub.org/moshannon/fctryfrm.htm.

Sierra Club. 2002(a). National Sewage Sludge Guidance. Circulated by email delivery; copy available from NEBRA or at http://www.sierraclub.org

Sierra Club. 2002(b). News Release: Sierra Club Asks EPA to Protect Communities From Toxic Sewage Sludge. July 3.

Similkameen Spotlight, The. 1992. Novel Use of Treated Wastewater Sludge Proposed To Help Reclaim Nearby Mine Sites And Promote Plant Growth On Rangeland. September 9.

Similkameen Spotlight, The. 1997. Green Grass Attracts Cattle. June 11.

Similkameen Spotlight, The. 1998(a). GVRD Reports Progress on Mine Reclamations. April 22.

Similkameen Spotlight, The. 1998(b). Mine reclamation projects on display as part of symposium. September 23.

Similkameen Spotlight, The. 2000. Mayer Inspects GVRD Project. October 17.

Slovic, P. 1993. Perceived Risk, Trust, and Democracy: A Systems Perspective. *Risk Analysis*, 13, 675-682.

Slovic, P. 1999. Trust, Emotion, Sex, Politics, and Science: Surveying the Risk-Assessment Battlefield. *Risk Analysis*, Vol. 19, #4.

Sludge. 1995. Washington Coalition Uses Education to Achieve Approval for Reuse Project. Vol. 20 Number 13. June 20.

Sludge. 2000(a). California OKs Land Application Rules that are More Time and Cost Efficient. 25(18):175.

Sludge. 2000(b). New Jersey Company Gets Approval to Apply Biosolids in Pennsylvania. 25(21):210.

Sludge. 2000(d). The Internet is Fueling the Debate. 25(8), 171. Business Publishers, Inc.

Sludge. 2002. EPA Inspector Report Does Little to Settle Biosolids Disputes. Silver Spring, MD: Business Publisher's, Inc., 40(7):68. April 8.

Smith, Bill. 1991. Indians Threaten Blockade To Stop Sludge Dumping. *Victoria Times Colonist.* February 20.

Smith, Gail. 1991. Sludge Seeks More Savory Image. *The Charlotte Observer.* April 17.

Snyder, David, and Frederick Kunkle. 2001. Health Fears Over Sludge Spur Quest for Controls: EPA Stand Challenged After Suspicious Deaths. *Washington Post.* B01. August 6.

Social Issues Research Centre, in partnership with the Royal Society and the Royal Institution of Great Britain. 2001. *Guidelines on Science and Health Communication.* November, pp. 1-8.

Solomon, John. (a). EPA Cites Sludge Recycling Concerns. *The Burlington Free Press,* February 7.

Solomon, John. 2002(b). Internal Probe Cites EPA for Lack of Research. *The Union Leader.* February 7.

Stark, S.A. 1993. Building Acceptance. *BioCycle.* 34(4):78-80. #2, important concept, land application, increasing public knowledge.

Stauber, John and S. Rampton. 1995. *Toxic Sludge is Good For You.* Monroe, ME: Common Courage Press.

Staudinger, Henry J. 1999. Land Application of Sludge (Biosolids), The Uncensored Story. A presentation to the Water Environment Federation seminar, Richmond, VA, November 4. http://www.safer-world.org/e/topics/sludge.htm .

Stevens, Jane E. 1998. Sludge-fest, Scientists Exchange Verbal Blows Over Risks of Recycled Sewage. *The Dallas Morning News*. March 23.

Stier, Jeff. 2000. Facts Versus Fears: How Science Does or Does Not Influence Public Opinion and Policy. *14th Annual Residuals and Biosolids Management Conference.* Water Environment Federation.

Stoddard, Carolyn L. 2001. Hill Man Hopes to be the First to Obtain Biosolids Under New Rules. *The Telegram*, 12(34), June 7-13. New Hampshire.

Stowe, Stacey. 1999. Dealing With Odor From Plant. *New York Times, Connecticut Weekly,* December 12, 1999.

Stukenberg, J.R., S. Carr, L.W. Jacobs, and S.B. 1993. *Final Report, Document Long-term Experience of Biosolids Land Application Programs*. Water Environment Research Foundation, Project 91-ISP-4.

Stulp, John R. 1995. Social, Political, and Educational Factors Involved in Facilitating Municipal Waste Utilization. In *Agricultural Utilization of Urban and Industrial By-Products,* American Society of Agronomy Special Publication no. 58, 1-9.

Sullivan, Dan. 1999. Toward Quality Biosolids Management: A Trainer's Manual.

Susskind, Lawrence. 1984. The Siting Puzzle: Balancing Economic and Environmental Gains and Losses. Presentation to the 13th Annual Conference of the Illinois Dept. of Environment and Natural Sciences, Sept. 13, 1984.

Susskind, Lawrence and Patrick Field. 1996. *Dealing with an Angry Public: A Mutual Gains Approach to Rebuilding Trust and Improving Long Term Relationships*. New York: The Free Press.

Susskind, Lawrence, Sarah McKearnan, and Jennifer Thomas-Larmer (eds.). 1999. *The Consensus Building Handbook: A Comprehensive Guide to Reaching Agreement.* Thousand Oaks, CA: Sage Publications.

Susskind, Lawrence, Paul Levy, and Jennifer Thomas-Larmer. 2000. *Negotiating Environmental Agreements: How to Avoid Escalating Confrontation, Needless Costs, and Unnecessary Litigation.* MIT-Harvard Public Disputes Program. Island Press.

Tackett, Stanford L. 1994. Op-Ed: The Sewage Sludge Scam. *The Indiana, PA Gazette,* October 2.

Talbot, Margaret. 2002. Hysteria Hysteria. *New York Times Magazine,* June 2.

Tannen, Deborah. 1990. You Just Don't Understand: Women and Men in Conversation. NY: Ballentine. pp. 281-298.

Thompson, Dan. 1995. Wetland Restoration with Biosolids and Yard Debris: Changing the Paradigm from NIMBY to YIMBY. *The Biosolids Trilogy*. Silverdale, WA. Northwest Biosolids Management Association 7th Annual Biosolids Management Conference.

Thompson, D.C., and B. True. 1994. The Future of Biosolids Recycling in the Pacific Northwest: Roadblocks, Roles and Responsibilities. In *12th Annual Residuals and Biosolids Management Conference*, Bellevue, WA: Water Environment Federation. 229-239.

Toffey, William E. 2000(c). Having Fun with Biosolids: Examining New Ways to Communicate the Benefits of Biosolids. Unpublished comments to Biosolids Tek-Con, Atlantic City, New Jersey, May 2000.

Toffey, William E. 2000(d). Virtuous Biosolids: Earning Love for Biosolids Through Our Commitment to Virtuous Practices. Presentation to the Mining, Forest, & Land Restoration Symposium/Workshop, July 17.

Toffey, William E. 2000(e). Latex or Mylar Biosolids Balloons: MABA's Experience in Getting the New Biosolids Organization to Float. Mid-Atlantic Biosolids Association presentation to Ohio WEA Conference, December 1, 2000.

Toffey, William E. 2002. Biosolids Good Practices: How Far Do We Go and How Do We Get There? Presentation to the Northwest Biosolids Management Association conference, September 23.

Touart, A.P. 1997. Creating An Effective Biosolids Information Network. *BioCycle*. August.

Touart, A.P. 1998. Acceptance Strategies: Winning Biosolids Support. *BioCycle* 39(2):86-90. February.

Trettin, L. and C. Musham. 2000. Is Trust a Realistic Goal of Environmental Risk Communication? *Environment & Behavior*, 32(3), 410-26.

Truini, Joe. 2001(a). NYC Drops Sludge Pact. *Waste News*. July 9.

Truini, Joe. 2001(b). Va. College Issues Compound Warning. *Waste News*, 7(7):13. August 6.

Tuler, Seth and Thomas Dietz (eds.). 2001. Factors Influencing the Success of Public Participation in Environmental Decision-Making: A Review of the Literature. July 10.

Tuohy, J. 2000. 2 Mothers, 2 Deaths, Too Many Questions—Sickness Was in the Air, But Officials Wouldn't Blame Sludge. *USA Today, Final Edition*. 13D. July 13.

U.S. Bureau of the Census, 2002, http://factfinder.census.gov

U.S. Environmental Protection Agency (U.S. EPA). 1977. *Municipal Sludge Management: Environmental Factors*. [MCD-28] EPA430/9-77-004, October.

U.S. EPA, 1980. *Information Programs Affect Attitudes Toward Sewage Sludge Use in Agriculture*. Municipal Environmental Research Laboratory, Cincinnati, OH: USEPA 600/2-80-103; July.

U.S. EPA. 1983(a). *Land Application of Municipal Sludge Process Design Manual.* EPA-625/1-83-016, October.

U.S. EPA. 1986. *Working to Gain Public Acceptance of Sewage Sludge Composting and Use of Liquid and Dewatered Sludge on Land*, U.S. EPA Office of Wastewater Management, March.

U.S. EPA. 1994(a). *A Plain English Guide to the EPA Part 503 Biosolids Rule*, Washington, D.C.: U.S. EPA Office of Wastewater Management.

U.S. EPA. 1994(b). *Biosolids Recycling: Beneficial Technology for a Better Environment.* Office of Water (4204), Washington, D.C.: U.S. EPA 832-R-94-009.

U.S. EPA. 1994(c). *Composting Yard Trimmings and Municipal Solid Waste*. Office of Solid Waste and Emergency Response, Report No. EPA/530/R94/003. 12-13, 108, 111-114.

U.S. EPA. 1995. *A Guide to the Biosolids Risk Assessments for the EPA Part 503 Rule*, Washington, D.C.: U.S. EPA Office of Wastewater Management.

U.S. EPA, Office of Solid Waste and Emergency Response. 1999. *Biosolids Generation, Use and Disposal in the United States*. EPA 530-R-99-009.

U.S. EPA Office of Inspector General. 2000. *Audit Report, Water, Biosolids Management and Enforcement (2000-P-10),* March 20.

U.S. EPA Office of Inspector General. 2002. Memorandum: EPA's Key Management Challenges. Washington, D.C.: U.S. EPA. September 6.

U.S. EPA, Office of Water, Office of Wastewater Management. 2000(a). *Guide to Field Storage of Biosolids.* EPA/832-B-00-007. July.

U.S. EPA, Office of Policy, Economics, and Innovation. 2001. *Stakeholder Involvement & Public Participation at the U.S. EPA: Lessons Learned, Barriers, & Innovative Approaches.* Washington, D.C. January.

U.S. EPA. 2002. Enhancing Facility-Community Relations. Brochure.

U.S. EPA, 2003b. Response to the Center for Food Safety et al. petition seeking an emergency moratorium… Washington, U.S. EPA Office of Water, December 24.

U.S. EPA, 2003c. Standards For The Use Or Disposal Of Sewage Sludge; Final Agency Response to the National Research Council Report on Biosolids Applied to Land and the Results of EPA's Review of Existing Sewage Sludge Regulations, in the *Federal Register,* December.

U.S. Geological Survey. Expanded Monitoring Program Near Deer Tail, Colorado. *USGS Quarterly Report*, 2(2):1-12, April-June 2000.

U.S. House of Representatives. 2000. Hearing Charter: EPA's Sludge Rule: Closed Minds or Open Debate? March 22, 2000.

UC Berkeley Wellness Letter. 2000. Organic Questions: The Answers May Surprise You. Berkeley, CA. October, p. 6.

U.S. District Court for the Western District of Virginia, Charlottesville. 1995. *W. Dale Welch, et al., Plaintiffs, v. The Board of Supervisors of Rappahannock County, Virginia, Defendant.* Civil Action No 94-002-C. May 24.

USA Today, 1999. Faced With Evidence of Health Threat, EPA Looks the Other Way. *USA Today* editorial, October 7.

USA Today. 2000. Faced With Faulty Science, EPA Muzzles Critics. *USA Today* editorial, Oct. 5. http://www.usatoday.com/news/comment/nceditf.htm

Vaughan, F., L. Koontz, and B. Barnes. 1994. Planning and Implementation of Wichita's Successful Biosolids Reuse Program. In *12th Annual Residuals and Biosolids Management Conference*, Bellevue, WA: Water Environment Federation. 347-353.

Vermont Public Interest Research Group (VPIRG). 1999. *On the Ground: The Spreading of Toxic Sludge in Vermont*. Montpelier, VT: Vermont Public Interest Research Group. June.

Virginia Cooperative Extension. 1999. *Agricultural Land Application of Biosolids in Virginia: Risks and Concerns.* Crop and Environmental Sciences Publication 452-304. Petersburg, VA: Virginia Cooperative Extension.

Virginia, State of. 2000. Biosolids Lifecycle: Concerns & Risks. http://www.biosolids.state.va.us/concerns.htm.

Wagner, E.O. 1990. Sludge Management Concerns in the United States. In *Water Pollution Control Federation Specialty Conference, The Status of Municipal Sludge Management for the 1990s*, New Orleans, LA: Water Pollution Control Federation. OGS-1–OGS-5.

Walker, John M., Robert W. Southworth, and Alan B. Rubin. 1997. EPA Regulations and Other Stakeholder Activities Affecting the Agricultural Use of By-Products and Wastes. Washington, D.C.: U.S. Environmental Protection Agency, Office of Water.

Wantland, Sheri (ed.). 2002. *Survival Guide: Public Communications for Water Professionals.* Alexandria, VA: Water Environment Federation.

Wardell, M.R. 1994. Building Acceptance for Biosolids Use. *BioCycle*. 35(7):56-59.

Washington State Department of Ecology. 2000. State biosolids recycling program. http://www.ecy.wa.gov/programs/swfa/biosolids/index.html.

Washington State University Cooperative Extension. 1998. Biosolids, A Recycled Organic Fertilizer. http://gardening.wsu.edu/stewardship/biosolid/biosolid.htm.

Washington, University of, Pack Experimental Forest. 1995. *Recycling a Resource: Biosolids Self-Guided Trail.*

Waste News. 2002. Land Application of Biosolids Draws Flak: EPA Neutralizes Stance. March 4.

Water Environment Federation and U.S. EPA. 1997. *Biosolids Fact Sheets*. March.

Water Environment Federation. 1990. *Biosolids—Too Valuable to Waste*. Brochure.

Water Environment Federation. 1994(c). *Biosolids Recycling—Beneficial Technology for a Better Environment.*

Water Environment Federation. 1994(d). Biosolids Recycling: Beneficial Technology for a Better Environment. Brochure.

Water Environment Federation. 1996. *Biosolids Public Acceptance Digest,* Water Environment Federation, Alexandria, VA.

Water Environment Federation. 1999(b). *Response To Biosolids Questions and Current Public Acceptance Issues*. WEF Residuals and Biosolids Committee, Public Acceptance Subcommittee.

Water Environment Federation. 2000(a). *Biosolids Success Stories.* Alexandria, VA: Water Environment Federation.

Water Environment Research Foundation. 2001. *Public Perception and Participation Workbook*, developed for WEFTEC 2001 conference, Atlanta, GA, October 14.

Water Environment Research Foundation. 2002. *The Development and Implementation of Successful Outreach and Community Partnering Programs*, a seminar workbook developed for WEFTEC 2002 conference, Chicago, IL, September 29.

West, R. 2000. Biosolids Carolina Style--Over Easy Without the Grits! *Beneficial Use of By-Products: Regional Issues,* a conference at The Ohio State University, December 1.

Wheelabrator Water Technologies. 2000. Wheelabrator Defendants' Memorandum of Law in Support of Motion to Exclude Plaintiffs' Expert Evidence on Causation. Filing at Rockingham Superior Court, NH.

Whitcomb, J.L., J.M. Halstead, L.C. Hamilton, and G.O. Estes. 1994. Community Issues in Facility Siting: The Case of Municipal Solid Waste Composting. New Hampshire Agriculture Extension Station, University of New Hampshire. 18p.

Wihtol, Christian. 2000. Tree Farm To Serve As Green Belt, Sludge Aid. *The Register-Guard.* July 3.

Wolz, Dan. 2000. The Bio-Chef and Living Creatively on the Other Side of the Nitrogen Cycle. *Beneficial Use of By-Products: Regional Issues,* a conference at The Ohio State University.

WASTEWATER UTILITY

Alabama
Montgomery Water Works & Sanitary Sewer Board

Alaska
Anchorage Water & Wastewater Utility

Arizona
Gila Resources
Glendale, City of, Utilities Department
Mesa, City of
Peoria, City of
Pima County Wastewater Management

Arkansas
Little Rock Wastewater Utility

California
Calaveras County Water District
Central Contra Costa Sanitary District
Contra Costa Water District
Crestline Sanitation District
Delta Diablo Sanitation District
Dublin San Ramon Services District
East Bay Dischargers Authority
East Bay Municipal Utility District
El Dorado Irrigation District
Fairfield-Suisun Sewer District
Fresno Department of Public Utilities
Irvine Ranch Water District
Las Virgenes Municipal Water District
Livermore, City of
Lodi, City of
Los Angeles, City of
Los Angeles County, Sanitation Districts of
Napa Sanitation District
Orange County Sanitation District
Palo Alto, City of
Riverside, City of
Sacramento Regional County Sanitation District
San Diego Metropolitan Wastewater Department, City of
San Francisco, City & County of
San Jose, City of
Santa Barbara, City of
Santa Rosa, City of
South Bayside System Authority
South Coast Water District
South Orange County Wastewater Authority
Sunnyvale, City of
Union Sanitary District
West Valley Sanitation District

Colorado
Boulder, City of
Colorado Springs Utilities
Littleton/Englewood Water Pollution Control Plant
Metro Wastewater Reclamation District, Denver

Connecticut
The Mattabasset District
New Haven, City of, WPCA

District of Columbia
District of Columbia Water & Sewer Authority

Florida
Broward, County of
Fort Lauderdale, City of
Gainesville Regional Utilities
JEA
Miami-Dade Water & Sewer Authority
Orange County Utilities Department
Orlando, City of
Reedy Creek Improvement District
Seminole County Environmental Services
St. Petersburg, City of
Stuart Public Utilities
Tallahassee, City of
Tampa, City of
Toho Water Authority
West Palm Beach, City of

Georgia
Atlanta Department of Watershed Management
Augusta, City of
Clayton County Water Authority
Cobb County Water System
Columbus Water Works
Fulton County
Gwinnett County Department of Public Utilities
Savannah, City of

Hawaii
Honolulu, City and County of

Illinois
American Bottoms Wastewater Treatment Plant
Greater Peoria Sanitary District
Kankakee River Metropolitan Agency
Metropolitan Water Reclamation District of Greater Chicago
North Shore Sanitary District
Wheaton Sanitary District

Iowa
Ames, City of
Cedar Rapids Wastewater Facility
Des Moines Metro Wastewater Reclamation Authority
Iowa City, City of

Kansas
Johnson County Unified Wastewater Districts
Unified Government of Wyandotte County/Kansas City, City of

Kentucky
Louisville & Jefferson County Metropolitan Sewer District

Louisiana
Sewerage & Water Board of New Orleans

Maine
Bangor, City of
Portland Water District

Maryland
Anne Arundel County Bureau of Utility Operations
Howard County Department of Public Works
Washington Suburban Sanitary Commission

Massachusetts
Boston Water & Sewer Commission
Upper Blackstone Water Pollution Abatement District

Michigan
Ann Arbor, City of
Detroit, City of
Holland Board of Public Works
Lansing, City of
Owosso Mid-Shiawassee County WWTP
Saginaw, City of
Wayne County Department of Environment
Wyoming, City of

Minnesota
Rochester, City of
Western Lake Superior Sanitary District

Missouri
Independence, City of
Kansas City Missouri Water Services Department
Little Blue Valley Sewer District

Nebraska
Lincoln Wastewater System

Nevada
Henderson, City of

New Jersey
Bergen County Utilities Authority
Passaic Valley Sewerage Commissioners

New York
New York City Department of Environmental Protection
Rockland County Solid Waste Management Authority/ Sewer District

North Carolina
Charlotte/Mecklenburg Utilities
Durham, City of
Metropolitan Sewerage District of Buncombe County
Orange Water & Sewer Authority

Ohio
Akron, City of
Butler County Department of Environmental Services
Columbus, City of
Metropolitan Sewer District of Greater Cincinnati
Northeast Ohio Regional Sewer District
Summit, County of

Oklahoma
Tulsa, City of

Oregon
Clean Water Services
Eugene/Springfield Water Pollution Control
Water Environment Services

Pennsylvania
Philadelphia, City of
University Area Joint Authority, State College

South Carolina
Charleston Commissioners of Public Works
Mount Pleasant Waterworks & Sewer Commission
Spartanburg Sanitary Sewer District

Tennessee
Cleveland, City of
Knoxville Utilities Board
Murfreesboro Water & Sewer Department
Nashville Metro Water Services

Texas
Austin, City of
Dallas Water Utilities
Denton, City of
El Paso Water Utilities
Fort Worth, City of
Gulf Coast Waste Disposal Authority
Houston, City of
San Antonio Water System
Trinity River Authority

Utah
Salt Lake City Corporation

Virginia
Alexandria Sanitation Authority
Arlington, County of
Fairfax County Virginia
Hampton Roads Sanitation District
Henrico, County of
Hopewell Regional Wastewater Treatment Facility
Loudoun County Sanitation Authority
Lynchburg Regional WWTP
Prince William County Service Authority
Richmond, City of
Rivanna Water & Sewer Authority

Washington
Edmonds, City of
Everett, City of
King County Department of Natural Resources
Seattle Public Utilities
Sunnyside, Port of
Yakima, City of

Wisconsin
Green Bay Metro Sewerage District
Madison Metropolitan Sewerage District
Milwaukee Metropolitan Sewerage District
Racine, City of
Sheboygan Regional Wastewater Treatment
Wausau Water Works

Australia
South Australian Water Corp.
Sydney Water Corp.
Water Corp. of Western Australia

Canada
Greater Vancouver Regional District
Toronto, City of, Ontario
Winnipeg, City of, Manitoba

Mexico
Servicios de Agua y Drenaje de Monterrey, I.P.D.

New Zealand
Watercare Services Limited

Singapore
Singapore Public Utilities Board

United Kingdom
Yorkshire Water Services Limited

STORMWATER UTILITY

California
Los Angeles, City of, Department of Public Works
Monterey, City of
San Francisco, City & County of
Santa Rosa, City of
Sunnyvale, City of

Colorado
Boulder, City of

Georgia
Griffin, City of

Iowa
Cedar Rapids Wastewater Facility
Des Moines Metro Wastewater Reclamation Authority

Kansas
Overland Park, City of

Kentucky
Louisville & Jefferson County Metropolitan Sewer District

Maine
Portland Water District

Minnesota
Western Lake Superior Sanitary District

North Carolina
Charlotte, City of, Stormwater Services

Pennsylvania
Philadelphia, City of

Tennessee
Chattanooga Storm Water Management

Washington
Seattle Public Utilities

STATE

Arkansas Department of Environmental Quality
Fresno Metropolitan Flood Control District, Calif.
Urban Drainage & Flood Control District, Colo.

CORPORATE

ADS Environmental Services
The ADVENT Group Inc.
Alan Plummer & Associates
Alden Research Laboratory
Aqua-Aerobic Systems, Inc.
Aquateam–Norwegian Water Technology Centre A/S
BaySaver Inc.
BioVir Laboratories, Inc.
Black & Veatch
Boyle Engineering Corporation
Brown & Caldwell
Burns & McDonnell
CABE Associates Inc.
The Cadmus Group
Camp Dresser & McKee Inc.
Carollo Engineers Inc.
Carpenter Environmental Associates Inc.
CDS Technologies Inc.
Chemtrac Systems Inc.
CH2M HILL
Damon S. Williams Associates, LLC
David L. Sheridan, P.C.
Dewling Associates, Inc.
Earth Tech Inc.
Eco-Matrix
Ecovation
Environmental Engineers International
EMA Inc.
The Eshelman Company Inc.
Finkbeiner, Pettis, & Strout Inc.
Freese & Nichols, Inc.
ftn Associates Inc.
Fuss & O'Neill, Inc.
Gannett Fleming Inc.
Golder Associates Inc.
Greeley and Hansen LLC
Hazen & Sawyer, P.C.
HDR Engineering Inc.
HNTB Corporation
HydroQual Inc.
Infilco Degremont Inc.
Ingersoll-Rand Energy Systems
Insituform Technologies Inc.
Institute for Environmental Technology & Industry, Korea
Jacobson Helgoth Consultants Inc.
Jason Consultants Inc.
Jordan, Jones, & Goulding Inc.
KCI Technologies Inc.
Kelly & Weaver, P.C.
Kennedy/Jenks Consultants
Komline Sanderson Engineering Corporation
Lawler, Matusky & Skelly Engineers, LLP
Limno-Tech Inc.
Lombardo Associates Inc.
Malcolm Pirnie Inc.
McKim & Creed
MEC Analytical Systems Inc.
Metcalf & Eddy Inc.
MWH
New England Organics
Odor & Corrosion Technology Consultants Inc. (OCTC)
Oswald Green, LLC
PA Government Services Inc.
Parametrix Inc.
Parsons
Post, Buckley, Schuh & Jernigan
R&D Engineering/Conestoga Rover & Associates
The RETEC Group
RMC, Inc. (Raines, Melton & Carella)
R.M. Towill Corporation
Ross & Associates Ltd.
Royce Technologies
SAIC Maritime Technical Group
Short Elliott Hendrickson, Inc.
Stantec Consulting Group, Inc.
Stormwater Management, Inc.
Synagro Technologies, Inc.
Tetra Tech Inc.
Trojan Technologies Inc.
URS Corporation
Veolia Water NATC
Wade-Trim Inc.
Weston Solutions Inc.
Woodard & Curran
WRc/D&B, LLC
WWETCO, LLC
Zoeller Pump Company

INDUSTRY

American Electric Power
ChevronTexaco Energy Research & Technology Company
The Coca-Cola Company
Dow Chemical Company
DuPont Company
Eastman Kodak Company
Eli Lilly & Company
Merck & Company Inc.
ONDEO Services
Procter & Gamble Company
PSEG Services Corp.
RWE Thames Water Plc
Severn Trent Services Inc.
United Water Services LLC

Note: List as of 7/1/04

WERF Product Order Form

As a benefit of joining the Water Environment Research Foundation, subscribers are entitled to receive one complimentary copy of all final reports and other products. Additional copies are available at cost (usually $10). To order your complimentary copy of a report, please write "free" in the unit price column. WERF keeps track of all orders. If the charge differs from what is shown here, we will call to confirm the total before processing.

Name ______________________ Title ______________________

Organization ______________________

Address ______________________

City __________ State __________ Zip Code __________ Country __________

Phone __________ Fax __________ Email __________

Stock #	Product	Quantity	Unit Price	Total
			Postage & Handling	
			VA Residents Add 4.5% Sales Tax	
			Canadian Residents Add 7% GST	
			TOTAL	

Method of Payment:

(All orders must be prepaid.)

- ❑ Check or Money Order Enclosed
- ❑ Visa ❑ Mastercard ❑ American Express

Account No. ______________________ Exp. Date __________

Signature ______________________

Shipping & Handling:

Amount of Order	United States	Canada & Mexico	All Others
Up to but not more than:	Add:	Add:	Add:
$20.00	$5.00*	$8.00	50% of amount
30.00	5.50	8.00	40% of amount
40.00	6.00	8.00	
50.00	6.50	14.00	
60.00	7.00	14.00	
80.00	8.00	14.00	
100.00	10.00	21.00	
150.00	12.50	28.00	
200.00	15.00	35.00	
More than $200.00	Add 20% of order	Add 20% of order	

*minimum amount for all orders

Note: Please make checks payable to the Water Environment Research Foundation.

To Order (Subscribers Only):

Log on to www.werf.org and click on the " Product Catalog."

Phone: (703) 684-2470
Fax: (703) 299-0742.

WERF
Attn: Subscriber Services
635 Slaters Lane
Alexandria, VA 22314-1177

To Order (Non-Subscribers):

Non-subscribers may be able to order WERF publications either through WEF (www.wef.org) or IWAP (www.iwapublishing.com). Visit WERF's website at www.werf.org for details.